Waste
Minimization
through
Process
Design

Other McGraw-Hill Environmental Engineering Books of Interest

AWWA • *Water Quality and Treatment*

BAKER • *Bioremediation*

BENNETT ET AL. • *Hazardous Materials Spills Handbook*

BRUNNER • *Hazardous Waste Incineration*

CHOPEY • *Environmental Engineering in the Process Plant*

CORBITT • *Standard Handbook of Environmental Engineering*

FREEMAN • *Hazardous Waste Minimization*

FREEMAN • *Standard Handbook of Hazardous Waste Treatment and Disposal*

HARRIS, HARVEY • *Hazardous Chemicals and the Right to Know*

JAIN • *Environmental Impact Assessment*

KOLLURU • *Environmental Strategies Handbook*

LEVIN, GEALT • *Biotreatment of Industrial and Hazardous Waste*

MAJUMDAR • *Regulatory Requirements for Hazardous Materials*

MCKENNA & CUNEO • *Pesticide Regulation Handbook*

SELDNER, COETHRAL • *Environmental Decision Making for Engineering and Business Managers*

SMALLWOOD • *Solvent Recovery Handbook*

WALDO, HINES • *Chemical Hazard Communication Guidebook*

WILLIG • *Environmental TQM*

Waste Minimization through Process Design

Alan P. Rossiter
Editor

McGraw-Hill, Inc.

New York   San Francisco   Washington, D.C.   Auckland   Bogotá
Caracas   Lisbon   London   Madrid   Mexico City   Milan
Montreal   New Delhi   San Juan   Singapore
Sydney   Tokyo   Toronto

628.4
WAS

Library of Congress Cataloging-in-Publication Data

Waste minimization through process design / Alan P. Rossiter, editor.
    p.   cm.
    ISBN 0-07-053957-X (acid-free)
    1. Factory and trade waste—Management.   2. Waste minimization.
I. Rossiter, Alan P.
TD897.847.W37   1995
628.4—dc20   94-49707
                                       CIP

1 2 3 4 5 6 7 8 9 0 DOC/DOC 9 0 0 9 8 7 6 5

ISBN 0-07-053957-X

The sponsoring editor for this book was Gail F. Nalven, the editing supervisor was Stephen M. Smith, and the production supervisor was Pamela A. Pelton. It was set in Century Schoolbook by Ron Painter of McGraw-Hill's Professional Book Group composition unit.

Printed and bound by R. R. Donnelley & Sons Company.

McGraw-Hill books are available at special quantity discounts to use as premiums and sales promotions, or for use in corporate training programs. For more information, please write to the Director of Special Sales, McGraw-Hill, Inc., 11 West 19th Street, New York, NY 10011. Or contact your local bookstore.

 This book is printed on recycled, acid-free paper containing a minimum of 50% recycled, de-inked fiber.

Passages in the Dedication taken from the *Holy Bible, New International Version®*. NIV®. Copyright © 1973, 1978, 1984 by the International Bible Society. Used by permission of Zondervan Publishing House. All rights reserved.

110485+-9

1/5/96.

To Belinda, Daniel, and Jason

"A wife of noble character who can find? She is worth far more than rubies." Prov. 31:10

"Sons are a heritage from the Lord, *children a reward from him."* Ps. 127:3

Contents

Part 3   Knowledge-Based Approaches

Part 4   Numerical Optimization Approaches

Part 5   Putting It into Action

Contributors

Richard J. Camm *President, Linnhoff March, Inc., Houston, Texas* (Chap. 3)

Vikas R. Dhole, Ph.D. *Linnhoff March, Ltd., Knutsford, England* (Chap. 7)

Prof. Thomas F. Edgar *Department of Chemical Engineering, The University of Texas, Austin, Texas* (Chap. 14)

Mahmoud M. El-Halwagi, Ph.D. *Department of Chemical Engineering, Auburn University, Auburn, Alabama* (Chaps. 15, 16)

Prof. L. T. Fan *Department of Chemical Engineering, Kansas State University, Manhattan, Kansas* (Chap. 13)

Darryl W. Hertz *Manager, Pollution Prevention Program, The M. W. Kellogg Company, Houston, Texas* (Chap. 19)

Y. L. Huang, Ph.D. *Department of Chemical Engineering, Wayne State University, Detroit, Michigan* (Chaps. 13, 14)

Jamil A. Khan, Ph.D., P.E. *Department of Mechanical Engineering, University of South Carolina, Columbia, South Carolina* (Chap. 11)

Howard Klee, Jr., Sc.D. *Director of Environmental Affairs and Safety, Amoco Corporation, Chicago, Illinois* (Chaps. 12, 21)

Jimmy D. Kumana *Linnhoff March, Inc., Houston, Texas* (Chaps. 4, 18, 23)

Prof. Bodo Linnhoff *Chairman and CEO, Linnhoff March International, Ltd., Manchester, England* (Chap. 5)

Debra Mitchell-Fox *Amoco Corporation, Chicago, Illinois* (Chap. 2)

Ravi Nath, Ph.D. *Setpoint, Inc., Houston, Texas* (Chap. 17)

Kenneth E. Nelson *President, KENTEC Incorporated, Baton Rouge, Louisiana* (Chaps. 10, 22)

Peter M. Nuttall, Ph.D. *Independent Environmental Consultant, Crewe, England* (Chap. 3)

Jerry J. Oliver *Vice President and Manager, Technology, Bechtel, Inc., Houston, Texas* (Chap. 1)

Michael K. Ozima *Southern California Edison Company, Rosemead, California* (Chap. 18)

Eric A. Petela *Linnhoff March, Ltd., Knutsford, England* (Chap. 6)

Mahesh K. Podar, Ph.D. *U.S. Environmental Protection Agency, Washington, D.C.* (Chap. 21)

Alan P. Rossiter, Ph.D. *Linnhoff March, Inc., Houston, Texas* (Chaps. 4, 9, 12, 17, 18, 23)

Ronald E. Schmitt   *Amoco Corporation, Chicago, Illinois* (Chap. 21)

Mohamed A. Serageldin, Ph.D.   *Triangle Speakers Group, Chapel Hill, North Carolina* (Chap. 20)

Richard L. Smith, Jr., Ph.D.   *Department of Chemical Engineering, Tohoku University, Sendai, Japan* (Chap. 11)

Robin Smith, Ph.D.   *Centre for Process Integration, University of Manchester Institute of Science and Technology, Manchester, England* (Chap. 8)

Debora M. Sparks   *Amoco Corporation, Cooper River, South Carolina* (Chap. 21)

Carol Stanley   *Department of Chemical Engineering, Auburn University, Auburn, Alabama* (Chap. 16)

Roger D. Sung, Ph.D.   *Southern California Edison Company, Rosemead, California* (Chap. 23)

Preface

The need to design and build "greener" plants has provided the great challenge for process engineering in the past decade. Increasingly, the focus has shifted from "end-of-the-pipe" solutions to more fundamental structural changes in industrial processes. These aim to eliminate emissions either by not creating pollutants in the first place or by recovering and re-using materials that would otherwise be discharged. This approach is known by many different names, including pollution prevention, waste minimization, and conservation-based environmental compliance. These terms are used virtually interchangeably in this book.

Part of the response to the environmental challenge has been the development of new systematic design methods that embody waste minimization concepts. These new approaches, most of which are based on process integration concepts, form the main subject matter of this volume. They are discussed in Parts 2 through 4. Several of the methodologies are dependent on software and, where appropriate, information is provided on relevant packages (shareware and/or commercial software). However, before addressing the methodologies we consider some of the issues—economic, practical, legal, and technical—that underlie current environmental trends. These are discussed in Part 1. Finally, in Part 5, we explore ways that the new approaches can be managed and implemented, and review several programs that embody these ideas to a greater or lesser extent.

Pollution prevention is a multifaceted, global problem. The authorship of the various chapters in this book reflects this fact. Individual contributors represent several different disciplines, notably chemical and mechanical engineering, law, and economics. Their backgrounds include process design, plant operation, government, consulting, academia, and business administration. Although most of the contributing authors are currently resident in either England or the United States, many have gained at least some of their professional experience in other parts of the world, notably Africa and Asia. The examples presented in this book are similarly diverse. Many industries are represented, notably chemicals, pulp and paper, metal plating, food and beverages, and dry cleaning; the oil and petrochemicals sectors are particularly prominent. Significantly, very similar approaches have proved useful in addressing waste minimization issues in all of these industries.

This book is written for professionals engaged in the design and operation of process plants. However, just as the problems of pollution prevention are diverse, so are the needs of the professionals who are active in this field. Some, especially those engaged in research and development, or in detailed design, require information with a great deal of technical depth. Others, particularly managers and decision-makers, are more likely to need a concise overview. In editing the material I have tried to balance these needs. Some technical details are included, but the reader who does not require this information can skip it without loss of continuity.

Acknowledgments

Many people have helped to make this book possible. Gail Nalven and the editorial staff at McGraw-Hill provided the initial impetus for the work, and helped to keep it on track. Each of the contributing authors had an important part to play. I am also grateful to my professional colleagues at Linnhoff March in England and the United States, and at Sasol in South Africa, for their support and encouragement. Special mention in this regard is due to Richard Camm, president of Linnhoff March, Inc. In addition, I am indebted to Duncan Woodcock of ICI Chemicals and Polymers for his helpful suggestions on parts of the manuscript; to Dennis Spriggs, formerly of Linnhoff March, for his help in defining the format and content of this book; and to Olive Ireland and Vrunda Dhole at UMIST for helping with manuscripts from England. Last, but by no means least, thanks are due to Kathleen Garza of Linnhoff March in Houston. She helped in many ways, ranging from typing to graphics to keeping me organized. Thank you one and all!

Alan P. Rossiter

Waste
Minimization
through
Process
Design

Setting the Stage

1

Pollution Prevention or Pollution Control?

Jerry J. Oliver
Vice President and Manager, Technology
Bechtel, Inc.
Houston, Texas

Introduction

Pollution is a global issue with real impacts on both the environment and human health. There are interminable debates on the absolute magnitude of these impacts and the best way to regulate them, but the simplest way to minimize pollution is not to produce it. This is the foundation for the pollution prevention concept. In contrast, most traditional approaches to managing pollution—and most current regulations—focus on controlling emissions at the "end of the pipe." These approaches have very different technical requirements and cost implications, with end-of-pipe control typically resulting in much higher operating costs. This will be illustrated in this chapter.

As the cost of handling hazardous and nonhazardous waste increases, it becomes easier to both visualize and quantify the extent of the problem or opportunity. An example is the exhaustive effort conducted by the National Petroleum Council (NPC) during the period 1991–1993 to estimate the cost impacts associated with environmental regulations for refineries in the United States. This study was the most thorough of its kind conducted by the industry. The key elements of the costs associated with refinery facilities provide a convenient method of focus on the significant and much broader issue of pollution prevention. The results of this study can be transferred to any industry segment in the United States as one attempts to aggregate the impacts of environmental compliance from a traditional end-of-pipe approach.

A powerful way to lessen the impact of the costs associated with compliance with environmental regulations is to use the *total quality management* (TQM) approach extended to include pollution prevention. As confirmed by a dozen demonstration projects conducted for the President's Commission on Environmental Quality (PCEQ), the same principles of TQM that helped to eliminate product defects or service deficiencies provide a systematic method to determine how to reduce pollution at a minimal cost. The TQM approach is illustrated in this chapter; other systematic methods for pollution prevention are described in later chapters.

Finally, lessons learned in the United States during the past few years can be readily extrapolated globally to develop new and upgraded facilities that are designed and operated in a fashion that minimizes or essentially eliminates pollution.

NPC Refinery Facilities Study on Environmental Impacts

Current environmental regulations for the process industry focus primarily on end-of-pipe control of pollution. In mid-1990, the Secretary of Energy requested that the NPC undertake a broad U.S. petroleum refining industry study to assess the current and perceived future impacts of these regulations between 1991 and 2010. The study was completed in August 1993[1] and utilized a broad industry task force, a panel of refining and environmental experts, an in-depth industry survey, and a highly qualified group of contractors and consultants. Four task groups were formed to handle various aspects of the broad effort: (1) industry survey; (2) refinery facilities; (3) refining capability and product quality; and (4) supply, demand, and logistics. The most relevant portion of the NPC study to this chapter is the refinery facilities section of the report, which includes the detailed development of costs associated with environmental compliance.

As background, the refinery facilities study involved the application of all current and potential environmental regulations to the entire U.S. refinery population, which in 1991 included 187 active and idled refineries. Sixty-one specific environmentally oriented line items were utilized to develop detailed cost information based on survey data and on current and future envisioned regulatory requirements. Investment and expense requirements for solutions to regulations were developed for each specific line item on a refinery group-by-group basis. The line items were aggregated into four broad categories: air, wastewater, solid waste, and health and safety.

The total investment and expense requirements developed were in excess of $150 billion with $36 billion in capital, $7 billion in one-time expenses, and $109 billion in incremental operating and maintenance (O&M) expenses (Fig. 1.1). Looking at each of the four categories in depth provides insight into the key cost areas anticipated. Tables 1.1 through 1.4 provide the key line items in each broad category. Capital requirements in the air sector are dominated by fugitives and particu-

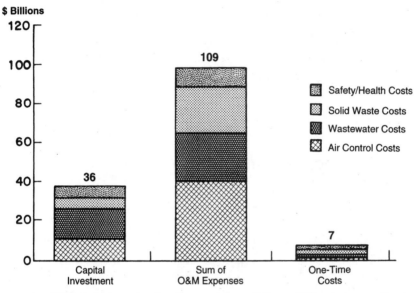

Figure 1.1 NPC study: impacts of environmental regulation on U.S. petroleum refineries from 1991 to 2010, total capital costs in 1990 dollars. (SOURCE: Ref. 1.)

TABLE 1.1   NPC Study: Impacts of Environmental Regulation on U.S. Petroleum Refineries from 1991 to 2010, Capital Cost Categories by Line Item

Air Sector

Line item	Billions*	Percentage of total
Fugitives, valves	$ 2.8	25
Fugitives, pumps	1.9	17
FCC particulate matter, redundancy	0.9	8
FCC particulate matter, new units	0.8	7
NO_x controls, heaters (SCR)	0.8	7
SO_x control, attainment—redundancy	0.6	5
SO_x control, attainment—new units	0.5	4
Coker handling	0.5	4
All others	2.5	23
	$11.3	100

*1990 dollars.
SOURCE: Ref. 1.

TABLE 1.2   NPC Study: Impacts of Environmental Regulation on U.S. Petroleum Refineries from 1991 to 2010, Capital Cost Categories by Line Item

Wastewater Sector

Line item	Billions*	Percentage of total
Two-stage ASP/PACT† units	$ 3.7	24
Incineration of sludge	2.8	18
Retrofit storage tanks—double bottoms	2.5	16
Store and treat storm water	1.4	9
Filtration of ASP/PACT effluent	1.2	8
Replace underground process piping	1.1	7
Process wastewater reuse	1.0	6
All others	1.9	12
	$15.6	100

*1990 dollars.
†Activated sludge plant/powdered activated carbon treatment.
SOURCE: Ref. 1.

TABLE 1.3   NPC Study: Impacts of Environmental Regulation on U.S. Petroleum Refineries from 1991 to 2010, Capital Cost Categories by Line Item

Hazardous and Nonhazardous Solid Waste Sector

Line item	Billions*	Percentage of total
Solid waste management units—inactive	$2.0	40
Aboveground tank replacement—light hydrocarbon	1.4	28
Aboveground tank replacement—heavy hydrocarbon	0.9	18
All others	0.7	14
	$5.0	100

*1990 dollars.
SOURCE: Ref. 1.

TABLE 1.4   NPC Study: Impacts of Environmental Regulation on U.S. Petroleum Refineries from 1991 to 2010, Capital Cost Categories by Line Item

Safety and Health Sector

Line item	Billions*	Percentage of total
Phaseout of hazardous material	$2.5	57
Process safety management	1.5	34
All others	0.4	9
	$4.4	100

*1990 dollars.
SOURCE: Ref. 1.

late matter controls followed by NO_x and SO_x capture. Wastewater, which is the largest capital cost category, is dominated by anticipated capital costs associated with anticipated *best available technology* (BAT) compliance requirements assumed to be part of the reauthorization of the Clean Water Act. Other major capital cost items in the wastewater sector involve retrofitting of storage tanks with double bottoms and treating or managing storm water. Solid waste capital requirements are dominated by closure of inactive solid waste management units and the replacement of aboveground tanks that are more than 40 years old with new double-bottom tanks. Finally, health and safety capital requirements are focused on the phaseout of hazardous materials and on process safety management. Table 1.5 provides a breakdown of capital costs by relative magnitude across all media which allows one to focus on the areas of potentially highest cost from both a waste management and a pollution prevention perspective.

Of the total costs, capital represents 25 percent while O&M expenses represent approximately 70 percent. This is typical of the end-of-pipe control solutions to pollution problems. However, O&M expenses can be reduced by the utilization of a pollution prevention approach on all capital projects at a refinery. Capital investments will be made regardless of the approach taken to compliance. Through a careful prioritization of each of the facilities' environmentally oriented capital programs, many of the required and discretionary capital expenditures can be made in a manner that both is inherently profitable and reduces waste volumes in critical areas. The path chosen will have a major impact on the refining industry during the next 20 years.

TABLE 1.5 NPC Study: Impacts of Environmental Regulation on U.S. Petroleum Refineries from 1991 to 2010, Capital Cost Categories by Line Item

Capital Cost Categories

Line item	Billions*	Percentage of total
Clean Water Act		
reauthorization—reduce toxicity	$ 7.9	22
Storage tank retrofits, double bottoms	4.8	13
Fugitives	5.1	14
Phaseout of hazardous material	2.5	7
Solid waste management unit—		
closures, retrofits, monitoring	2.2	6
Particulate matter controls	2.2	6
Process safety management	1.5	4
All others	10.1	28
	$36.3	100

*1990 dollars.
SOURCE: Ref. 1.

This detailed case study on a straight end-of-pipe approach to compliance is reproducible across industries and represents the impacts that will be seen if a pollution prevention concept to waste management is not applied. The impact of using the pollution prevention approach is substantial by reducing both capital and incremental operating and maintenance costs and by potentially increasing and improving production.

Current Situation

It is useful to look at where we are today from an environmental perspective. Awareness of environmental impacts continues to increase, putting broad pressure on environmental issues. This focus is on the individual, the facility, and the industry in a balanced fashion. The cost of remediating problems in the United States continues to rise, making environmentally driven projects the largest part of many companies' capital budget and a large portion of allocated O&M expense. Regulations are increasing in the United States and globally. Emphasis is being placed on voluntary efforts by industry, but command and control programs are continuing in the United States and developing internationally, especially when voluntary programs do not lead to the desired results. End-of-pipe solutions to environmental regulations and problems are reactionary and do not allow for the most cost-effective solutions. Further, as exhibited by the AMOCO/EPA effort completed in late 1992 at Amoco's Yorktown Refinery (see Chap. 21), a procedural approach to compliance in the reactionary mode sometimes ignores the real source of the problem.

To bridge the current perceptions in a proactive fashion and to move away from the end-of-pipe reactionary approach, I advocate the application of the TQM approach to the pollution prevention concept.

Pollution Prevention

The prevention of waste will be the environmental focus for the next 10 to 20 years. The concept has been called *pollution prevention, source reduction,* and *waste minimization,* but all have the same connotation of reducing waste. The idea is that it is better not to produce waste in the first place than to have to worry about treatment and disposal. Pollution prevention seems to have a broader connotation than the other terms used to describe a reduction of waste and is currently a term broadly used throughout industry and government. Keep in mind that the concept is not new. In fact, the old refinery industry example is worthwhile for perspective.

In the late 1800s, the refining industry was born on the banks of the Cuyahoga River in Ohio with hundreds of pot still refineries to recover kerosene from crude oil. Everything else, including gasoline, was dumped into the river. This activity created the image of pollution for Cleveland that lives today as "the river that burned." The refinery industry recognized the folly of wasting such a large percentage of the barrel of crude oil and began a steady dedicated progression of development of products from the crude oil waste until the present, when almost all the barrel is successfully converted to usable products.

In order to apply the simple pollution prevention precept, three things are needed and were applied to the example above:

1. Belief in the concept that waste can be eliminated profitably

2. Technological innovation

3. Management commitment

A recent example of the positive effects of applying the pollution prevention approach to a regulation-driven problem is provided in a paper given by Bechtel and Union Carbide at the 1992 Project Management Institute seminar/symposium in Pittsburgh.[2] The problem involved the need to treat wastewater from Union Carbide's petrochemical complex at Texas City, Texas. The need was driven by the Clean Air Act under the pretreatment standards for existing sources (PSESs). Shortly after the regulations were promulgated, an environmental consultant was hired to conduct a sampling program using traditional end-of-pipe techniques. Based on this sampling program and the baseline monitoring report that followed, a biological treatment facility was specified as the proper approach to compliance. The need for total capture necessitated two biological treatment trains with one train being redundant. The cost of this fix was estimated to be $14 million. The work was put out for bid, and Bechtel won the work under its alliance with Union Carbide. Based on a review of the data, it was determined that many of the constituents that were to be captured in the biological treatment facility were not present in the plant. Further, because the entire effort to that point had been at the end of the pipe, a plant material balance was not included. When the collected data were reviewed, a material balance was completed, and the interconnections of the biological plant were considered, a few conclusions were obvious. Many of the constituents that were to be captured in the biological treatment plant were not present in the facility (only 8 out of 40 were actually present). The design basis for the biological treatment facility was rendered invalid, and a design basis for a plant that would fit all end-of-pipe and in-plant data led to a project cost of $45 million for a similar scope of work. Union Carbide canceled the project. Unfortunately the regulatory need still existed, and the period of compliance was now down to 1 year. Union Carbide and Bechtel moved away from an end-of-pipe focus and examined the entire plant in light of the regulatory need. Nothing was considered sacred in this review. The team that was assembled was able to identify six major sources of pollution and was able to develop solutions for each. Some of the solutions were as easy as reconfiguring piping and replacing sumps, and one of the solutions led to the installation of a stripper to remove aromatics at the source. The final result was that the entire job was done for $15 million, was done on time, and did not require any end-of-pipe treatment.

A key precept in driving industry toward a pollution prevention approach is that pollution prevention pays (see Chap. 22). The above example and the next section of this chapter provide concrete evidence that compared to the end-of-pipe or do-nothing alternatives, pollution prevention results in significant cost savings. The next example deals with ensuring that a pollution prevention project is analyzed in total versus its alternative. If a thorough examination is made of the positive impacts, a compelling argument on the value of pollution prevention can be developed. This example is from the July–September 1993 *EPA Journal*[3] and applies to the pulp and paper industry. The Environmental Protection Agency (EPA) analyzed the economics of a pollution prevention project at a fine paper mill that was designed to reuse fiber, filler, and water on two paper machines on a continuous basis. This industry is being pressured to reduce overall waste, and the concept is to work on projects that reduce waste while improving overall plant efficiency and conserving both raw materials and utilities. Before this project was undertaken, waste material generated by the paper machines was handled in the plant's wastewater treatment system. The company made a $1.5 million investment in the project. The company's estimate on the value of the project indicated that annual savings of $0.36 million would be realized. This yields in an *internal rate of return* (IRR) of 21 percent (based on a 15-year plant life), and a simple payout of 4.2 years would be realized. The analysis done included the obvious savings involved with less waste handling and raw material reductions but did not consider the savings associated with freshwater pumping, treatment, and heating as well as wastewater pumping. When these savings were included with the obvious savings, the IRR increased to 48 percent, the payout shortened to 1.6 years, and the positive cash flow increased to $2.85 million over a 15-year period. The improved economics when all plant efficiency gains were included were significant enough to move a marginal project into the realm of a worthwhile capital investment without worry about the environmental drivers.

A TQM Approach to Pollution Prevention

End-of-pipe or do-nothing solutions to environmental problems may address many regulatory requirements, but in general they offer no other benefits to the company that implements them. One of the best ways to ensure that the positive impacts of environmental projects are maximized is by utilizing a total quality management approach with an emphasis on pollution prevention. The following elements apply:

- The definition of quality is expanded to include the environmental impacts of products and operations.
- The approach is applied consistently.
- The belief exists that like quality, pollution prevention pays.
- Continuous improvement is needed.
- Training and education are ongoing.
- Teaming alliances and external input remove paradigms.

To provide a quantifiable example of the impact of using a TQM approach to pollution prevention, a brief look at a program under the President's Commission on Environmental Quality (PCEQ), is worthwhile.[4] In 1991, President Bush put together the PCEQ, composed of 25 business, environmental, foundation, and academic leaders. The commission worked for 18 months, in cooperation with the government, on initiatives for environmental improvement. One of the main initiatives involved the application of TQM to the pollution prevention process. This was accomplished by setting up a dozen diverse demonstration projects. Each project exhibited significant pollution prevention and economic savings. In aggregate, the projects accomplished the following:

- Eliminated thousands of tons of pollutants
- Saved substantial sums of money
- Increased plant efficiency and effectiveness
- Improved the quality of both products and services
- Enhanced public perception
- Improved employee morale

The following projects illustrate the successes of the program:

1. Ford's Climate Control Division plant, in Plymouth, Michigan, successfully conducted a pilot project to replace a toxic chemical, trichloroethylene (TCE), with an aqueous detergent in the degreasing of aluminum heat exchangers. They got a superior product, reduced costs and environmental impacts, and improved health and safety conditions. The project is now a model for incorporating TQM into product design and manufacturing processes.

2. GE Medical Systems, in Florence, South Carolina, challenged a self-directed workforce to integrate environmental concerns into its

existing continuous improvement processes, which resulted in a variety of projects that have improved both their environmental and economic performance.

3. International Paper's Androscoggin Mill in Jay, Maine, worked with various environmental and government organizations to design and implement a quality environmental program, resulting in improved stakeholder relations and reduced environmental impacts. The *biological oxygen demand* (BOD) to the Androscoggin River was reduced 65 percent, and fiber loss in the mill sewers is down 50 percent.

4. Dow Chemical's Glycol II plant, in Plaquemine, Louisiana, used internal benchmarking to achieve a 29 percent reduction in fugitive emissions of ethylene oxide. The plant strove to be "leak-free by 1993." By using simple tools of quality performance, Dow Chemical also reduced its laboratory waste by 67 percent.

5. 3M's Medical Products Division facility, in Brookings, South Dakota, integrated pollution prevention throughout its entire operation, engaging every employee in pollution prevention activities. This resulted in a 10 percent reduction in waste generated. 3M has had an enviable program called *Pollution Prevention Pays* since 1975, with documented savings of more than $500 million. The application of TQM to the program improved results even further.

6. Chevron's Perth Amboy refinery, in Perth Amboy, New Jersey, developed a facilitywide pollution prevention initiative to reduce environmental costs and improve competitiveness. Cross-functional teams identified more than 300 opportunities for improvement and narrowed the list to 10 for immediate action.

7. Du Pont's acrylonitrile operations, in Beaumont, Texas, made process modifications that reduced ammonium sulfate waste from more than 100 million lb/yr to less than 40 million lb/yr. What surprised the Du Pont team was that Du Pont set out to reduce deepwell disposal volumes even if it cost them money. What resulted was reduced raw material purchases and lower disposal costs with little capital investment. Du Pont ended up saving almost $1 million per year in manufacturing costs with essentially no capital investment.

8. Procter and Gamble's paper products plant in Mehoonpany, Pennsylvania, which makes Pampers, Luvs, and Charmin bathroom tissue, identified priority waste streams by using TQM techniques and developed a comprehensive strategy to reduce those wastes. Savings of $500,000 per year are projected for the pollution prevention projects identified during the PCEQ program.

9. Merck's Rahway manufacturing facility, in Rahway, New Jersey, focused on the reduction of SARA toxic release inventory emissions, and it achieved a 1.5 million lb reduction in these releases in 1992 and projected a 4.8 million lb reduction in 1993. Merck has targeted a 90

percent reduction by 1995 from a 1987 base. The use of TQM was viewed favorably as having accelerated progress toward the 1995 goal.

10. Finally, Bechtel contributed a project through our (partnership of Pacific Gas & Electric) U.S. Generating Company's Indiantown cogeneration facility, in Martin County, Florida, that demonstrated the impact of utilizing quality principles to incorporate environmental considerations into the siting, development, construction, and operation of a clean-coal facility. U.S. Generating's proactive commitment to state-of-the-art technologies resulted in a streamlined permitting process that expedited financing and allowed for a prompt construction start.

All the above projects are unique in two important respects. First, several demonstrations directly involved national and community environmental public interest groups. Second, the participating companies successfully showed that improvements in environmental quality can be achieved by using the same TQM principles that have been applied by companies worldwide to give their customers better products and services at lower cost. It was strongly felt by the participants that these demonstration project results were indicative of the potential pollution reduction that might be achieved throughout industry.

Projects that are driven by management attention, by employee dedication, and by simple low-cost changes in procedure or practice are being worked on in a global fashion. By applying the broad principles of TQM, a dedicated continuous improvement process is put into place and provides a basis for an ongoing reduction of waste. Once the easy-to-solve and low-capital-investment projects are completed, the next increment of projects to reduce waste may require significant capital investments, but could be profitable projects in their own right. The key to a successful TQM-driven pollution prevention program is that as much waste reduction as possible is wrapped into positive cash flow projects. These projects must be planned in a positive, proactive manner and not driven by an end-of-pipe reaction to environmental regulations.

Financial Considerations

With the number of capital projects that are necessary simply for competitive and quality purposes, environmentally driven projects tax a capital-constrained industry. If all capital projects are pursued in a manner that provides a pollution prevention focus, then it is feasible that, as discussed above, many of those projects will be profitable and will serve dual purposes of increasing competitiveness and eliminating pollution.

With the creative financing programs that have been developed for the *independent power-producing* (IPP) industry, a significant number of new off-balance-sheet financial tools exist that can allow a capital-constrained but otherwise strong industry to invest capital without significantly increasing debt. This involves the use of outside investors who may or may not have an interest in the business over the long term, but are interested in the rewards present in the investment from both a debt and an equity perspective. Many of the more sophisticated equity investors are willing to take sufficient market-driven risks to allow the investment to be truly off balance sheet and outside the normal internally raised sources of equity. The concept is normally called *build, own, operate* (BOO) or *build, own, transfer* (BOT). It is as applicable now to the process industries as it has been to the IPP industry during the past several years.

Conclusions

Environmental awareness is increasing globally, and the cost implications of "business as usual" are being quantified. The philosophy and approach to solving problems that are raised by this increasing environmental awareness will make the difference in how business is done in the future. The right approach will unleash the power to convert the problems to opportunities. Methods exist to extract greater value from suppliers. There is more focus on customer needs than ever before. In summary:

- Based on the work of the NPC, the cost of compliance with environmental regulations (which mostly encourage end-of-pipe pollution control) between 1991 and 2010 is in excess of $150 billion. The capital component of that cost is $36 billion.

- Pollution prevention is a viable way to lessen the impacts of current and developing environmental regulations in a proactive manner. The concept is being broadly embraced by both industry and government.

- Pollution prevention pays. To demonstrate this, all the savings must be documented and then the message of those savings versus the end-of-pipe or do-nothing alternatives distributed widely.

- Utilizing a TQM approach improves quality and leads to a greater impact of the pollution prevention effort.

- Alternatives to traditional approaches to capital raising are evolving and should be used to maximize the benefits of pollution prevention.

Finally, some solutions and technologies will be shared more effectively, and others may be held for a real competitive edge; barriers that contributed to the rise of untenable regulations can be overcome; and resource constraints are yielding to innovation.

References

1. *U.S. Petroleum Refining—Meeting Requirements for Cleaner Fuels and Refineries,* National Petroleum Council, Washington, 1993.
2. B. P. Pohani and E. C. Brod, *Quality Management of Environmental Compliance Projects and Waste Minimization,* PMI, Pittsburgh, 1992.
3. A. L. White, "Accounting for Pollution Prevention," *EPA Journal,* Washington, 1993.
4. *Total Quality Management—A Framework for Pollution Prevention,* President's Commission on Environmental Quality, Washington, 1993.

2

Legal Aspects of Pollution Prevention

Debra Mitchell-Fox
Amoco Corporation
Chicago, Illinois

Environmental protection is a subject of global concern. In the United States, the regulation of pollution began in 1899 with the Rivers and Harbors Act, which was designed to protect navigation from surface water discharges. The current era of environmental regulation began in 1963 with the adoption of the Clean Air Act. Since then, numerous environmental statutes have been adopted and amended, and thousands of pages of environmental regulations have been issued. Over the last 30 years, environmental regulations have become much more detailed and complex. Federal environmental regulations currently span over 8000 pages.

Complex environmental regulations exist and continue to evolve in many countries, and it is not possible to address all these here. This chapter focuses primarily on current environmental regulations in the United States and on international issues. It includes an overview of the three primary federal environmental statutes—the Clean Air Act (CAA), Clean Water Act (CWA), and the Resource Conservation and Recovery Act (RCRA)—and their corresponding regulations, and it also addresses some of the international environmental agreements that have become increasingly important in recent years. It will begin with a discussion of the legal framework for adopting environmental laws and will close with a summary of the implications for pollution prevention design and implementation.

Legal Framework

Environmental statutes are adopted at the federal level by Congress. Statutes typically establish general guidelines, although environmental statutes have become very detailed and prescriptive over the last several years.

The Environmental Protection Agency (EPA) is the primary agency for implementing environmental statutes at the federal level. The EPA adopts detailed regulations in order to implement statutory provisions. The EPA must publish proposed regulations in the *Federal Register* and accept comments from interested parties prior to finalizing a rule. The rule-making process can often take 1 year or more to complete, from initial conception by the EPA through publication of the final rule in the *Federal Register*. Once a regulation is finalized, it is incorporated into the *Code of Federal Regulations,* or *CFR,* which is a compilation of federal regulations.

Statutes and regulations adopted at the federal level apply throughout the country and do not necessarily address site-specific situations. It may not be clear, e.g., whether a permit is required for a particular construction project. Consequently, interpretation of environmental laws by an expert may often be required. Such experts will rely on the formally adopted statutes and regulations as well as other less formal sources of information. For example, the EPA publishes guidance documents and issues memorandums which are not legally binding but provide guidance regarding the applicability of a particular regulatory provision. Proposed and final regulations published in the *Federal Register* are typically accompanied by a lengthy preamble, which is essentially the EPA's discussion of the contents of the rule and may provide guidance regarding the rule's intent.

At the state level, state legislatures adopt state environmental statutes, and state agencies such as the Texas Natural Resource Conservation Commission (TNRCC) or the Indiana Department of Environmental Management (IDEM) adopt regulations.

As a general rule, the authority to implement *federal* environmental programs is delegated to the states. However, a state may have authority to implement only the Clean Air program, e.g., but not the Clean Water or RCRA programs, or a state may have authority to implement only a portion of one of the programs. Consequently, both state and federal regulations often apply to a particular facility. While federal environmental laws typically serve as a model for state laws, environmental laws differ from state to state. As a result, the United States has a labyrinth of environmental laws, and the requirements applicable to each specific project should be investigated. This chapter summarizes the major federal environmental laws, but state laws may have additional or different requirements.

Major Federal Environmental Laws

Clean Air Act

The Clean Air Act (CAA) was adopted in 1963 and has been amended on numerous occasions, including the most recent amendment in 1990. The three primary programs under the CAA regulate (1) the emission of so-called criteria pollutants, (2) the emission of hazardous air pollutants, and (3) mobile source emissions.

Criteria pollutants. The EPA established *national ambient air quality standards* (NAAQSs) for six criteria pollutants: sulfur dioxide (SO_2), ozone (O_3), carbon monoxide (CO), lead (Pb), nitrogen dioxide (NO_2), and respirable particulate matter (PM_{10}). The EPA also divided the nation into approximately 250 air quality control regions. Each region is then rated as either in attainment or in nonattainment with the ambient air quality standards for each of the six criteria pollutants. For example, a region may be classified as attainment for lead and nonattainment for ozone. Under the 1990 amendments, all regions must achieve attainment with all criteria pollutants by specified dates, ranging from 1993 to 2010.

Each state develops a *state implementation plan* (SIP) for achieving and maintaining attainment with the NAAQSs. SIPs must meet certain federal requirements, but the state may choose its own mix of emission limitations and control measures. For example, one state may choose to have more stringent mobile source controls and less stringent stationary source controls for *volatile organic compounds* (VOCs). In general, a separate regulation will be adopted for each of the six criteria pollutants, although there may also be specific requirements for certain categories of facilities.

Hazardous air pollutants. Prior to 1990, the EPA issued regulations for only eight *hazardous air pollutants* (HAPs). Congress became impatient with the EPA and identified 189 HAPs in the 1990 Clean Air Act Amendments. The EPA is currently in the process of developing regulations for these newly identified HAPs. Regulations will generally be adopted for each industrial category, such as pulp and paper, refining and semiconductor manufacturing, rather than for each pollutant. The categorical standards will generally specify the technologies which will meet the statutory control criteria, referred to as the *maximum achievable control technology* (MACT).

Mobile sources. There are special provisions in the CAA related to mobile sources such as automobiles, boats, and lawn mowers. For example, the CAA contains tailpipe standards for motor vehicles and requirements for motor vehicle fuels. Under the 1990 amendments, special fuels must be used in certain nonattainment areas. For example, reformulated gas must be used in nine ozone nonattainment areas by 1995, and oxygenated fuels have been used in 41 carbon monoxide nonattainment areas since November 1992.

Technology requirements. The CAA has a clear preference for demonstrated, versus innovative, technologies. New sources in nonattainment areas must install control devices which will achieve the *lowest achievable emission rate* (LAER). New sources in attainment areas must install the *best achievable control technology* (BACT), and sources with HAP emissions must install the maximum achievable control technology. The LAER (for nonattainment areas), BACT (for attainment areas), and MACT (for HAPs) are all defined in the statute in terms of "emission limitations" or "emission reductions." However, in practice, the standards correspond to a particular technology which has demonstrated a certain removal efficiency or emission reduction percentage.

For example, 98 percent destruction efficiency may be considered BACT for hydrocarbon vent control in a particular state. While this appears to be a performance standard, in reality it may be the same as specifying combustion as the control technology. The standard represents what is technically achievable by using the best available control technology. Unfortunately, this preference for demonstrated technologies may limit a facility's flexibility and pollution prevention options.

If hydrocarbons are recycled or recovered by being used as a fuel, e.g., recovery may achieve only 95 percent destruction efficiency. The 98 percent BACT solution (combustion) would result in the emission of additional NO_x, whereas the recycling option would not. Nevertheless, many states may not allow a deviation from the 98 percent BACT standard, even where the net environmental benefits associated with recovery are greater.

The EPA and the states have the authority to allow flexibility and encourage pollution prevention options in some situations, although they do not always take advantage of these opportunities. The EPA recently adopted several policies which attempt to promote additional flexibility. For example, the EPA advocates greater emission trading among emission units and between mobile and stationary sources. The EPA also advocates the use of market incentives such as tradable permits, which allow a facility to save unused emission allowances for future use or to sell or trade their unused allowances to other sources.

Permitting. Any major new source or major modification to an existing source of emissions must have a permit prior to construction. This process is referred to as *new source review* (NSR). In a nonattainment area, a facility must identify sufficient offsetting emission reductions to counterbalance the project's new emissions. The new source must also install the lowest achievable emission rate achieved by demonstrated technologies at other facilities.

A different set of NSR regulations apply in attainment areas. The regulations are designed to prevent significant deterioration of existing air quality and are therefore referred to as *prevention of significant deterioration* (PSD) regulations. New emissions may only "deteriorate" or affect the ambient air quality by a specified increment. In addition, a new source must install the best achievable control technology.

Pursuant to the 1990 amendments, a new permit program will be implemented over the next several years. Currently, a complex facility may have 20 or 30 air permits, one for each emission unit or group of units within the facility. Under the new permitting program, each facility will have one consolidated permit.

Pollution prevention. Before a facility can construct a pollution prevention project, it must determine whether the project constitutes a "major modification." If so, the project is subject to the new source preconstruction review and permit process. A *major modification* is defined as a physical or operational change which will result in a significant net emissions increase. However, because of complex regulatory definitions, pollution prevention projects which will result in a net *reduction* in actual emissions could also require a permit prior to construction.

Since the NSR process occurs prior to construction, the facility must predict future emissions. Because of the corresponding uncertainty, NSR regulations provide that future emissions will be presumed to equal the unit's potential to emit at full capacity. Consequently, the facility effectively compares current actual emissions (generally at less than full operating capacity) with future potential emissions (assuming full operating capacity). As a result, future emissions may be over-estimated, and many projects will trigger NSR even though they will actually reduce plant emissions.

For example, assume that a facility has historically operated at 75 percent capacity and wishes to switch from coal to natural gas. Also assume that a physical change is necessary in order to achieve the switch, but the facility intends to continue operating at the same production rate. Actual emissions will decrease as the unit switches to cleaner-burning natural gas. However, the facility must compare current actual emissions while using coal at 75 percent capacity with projected future emissions while using natural gas at 100 percent capacity. Such a comparison may very well predict a *significant net emissions increase,* as the term is currently defined in NSR regulations.

In 1992, the EPA amended the NSR regulations to allow electric utilities to compare "actual to actual" emissions in certain situations. This rule is generally referred to as the *Wepco rule* after a court case brought by the Wisconsin Electric Power Company. It effectively means that a permit is not required for projects which result in net emission reductions. However, the rule has not yet been expanded to other industrial categories.

Clean Water Act

The Clean Water Act was adopted in 1972, and it regulates the discharge of pollutants into surface waters. The requirements are implemented through the *national pollutant discharge elimination permitting system* (NPDES). A permit is required for point-source discharges and will generally specify the amount of pollution that can be discharged at the end of a pipe, based on technological guidelines and water quality standards.

Technology-based effluent standards. The EPA established technology-based effluent guidelines for several categories of facilities such as synthetic organic chemical manufacturing. These categorical standards prescribe effluent concentrations for conventional, nonconventional, and toxic pollutants. Conventional pollutants include *biological oxygen demand* (BOD), *total suspended solids* (TSS), pH, and fecal coliform. The EPA identified over 60 toxic pollutants, including benzene, copper, and mercury. Nonconventional pollutants are pollutants other than conventional or toxic pollutants, such as iron and ammonia.

The technology-based effluent standards are derived from an assessment of what can be achieved with existing, demonstrated technologies. The categorical standards are applied to individual facilities through the NPDES permitting program. Effluent standards specified in a permit are subject to the antibacksliding rule, which says that effluent limitations cannot become less stringent in subsequent permits or permit amendments.

Water-quality-based effluent standards. Water quality standards are established based on the designated uses for a receiving water body, such as a public water supply, recreational or industrial. The antidegradation policy provides that existing water quality cannot be degraded by a discharge (except in rare circumstances), even if the existing water quality exceeds what is necessary to support the designated use.

Where technology-based effluent limitations established in a permit will not achieve the water quality standards, the permit will establish more stringent, water-quality-based effluent limitations designed to meet the water quality standards. Once water-quality-based effluent limitations are established in a permit, the antibacksliding rule will preclude them from being relaxed in the future.

Pollution prevention. Effluent limitations under the CWA tend to focus on end-of-pipe treatment rather than pollution prevention. In some cases, a source reduction project may result in a greater net environmental benefit, but may not meet certain effluent limitations or may require backsliding of permit terms. For example, assume that an existing activated sludge plant provides secondary treatment which achieves a 99 percent soluble organic removal efficiency. Assume also that the treatment plant meets all effluent standards for toxic materials but cannot consistently meet the TSS standard. The TSS standard was presumably established based on what is technically achievable, not necessarily on what is necessary to protect human health and the environment.

The facility undertakes an extensive research and development project to attempt to meet the TSS requirement, exploring filtration, carbon adsorption, pretreatment, system reconfiguration, and other technological options. The only solution which will meet the TSS effluent limitation involves tertiary treatment, an additional activated sludge polishing step. However, removing TSS in a tertiary treatment system results in the generation of additional solids which will have to be deposited in a landfill or land farm.

The facility also looks at a source reduction option. The source reduction project will involve twice the initial capital expenditure, but will result in a positive net present value due to the recovery of product and treatment plant operational cost savings. Nevertheless, while the source reduction option will reduce TSS, it will still not quite meet the TSS effluent limit on a consistent basis. Therefore, under current permitting rules, the source reduction option would not satisfy the regulatory requirements, even though it may result in a greater net environmental benefit.

Resource Conservation and Recovery Act

Hazardous waste. The RCRA is designed to regulate hazardous waste from "cradle to grave" by regulating generators, transporters, and disposers of hazardous waste. A generator must determine if a particular material is a solid waste within the meaning of RCRA and, if so, if the material is hazardous. If the material is a hazardous waste, then it is subject to the full range of RCRA regulations.

A *solid waste* is defined in the statute as any discarded material which could include liquids, sludges, and contained gases, as well as solids. This seems rather straightforward, but there are a number of exemptions and exclusions to the definition and it is very complicated in practice. For example, experts often differ as to whether a material is actually "being discarded," particularly where it is reused or recycled in certain ways. Certain process treatment units, such as a sour water stripper, may be considered a process unit from an engineering standpoint, but may be considered a waste treatment unit from a regulatory standpoint.

If a material is a solid waste, it may be considered hazardous in one of two ways. First, the EPA has defined a number of "listed hazardous wastes" from specific industries, such as lead smelting and iron and steel manufacturing, which are presumed hazardous. If a solid waste is not listed as hazardous, it may be hazardous due to one of four characteristics: toxicity, ignitability, corrosivity, or reactivity.

In addition to being listed as a hazardous waste or characteristically hazardous, a waste may be regulated as hazardous under either the "mixture" or "derived from" rule. If a nonhazardous waste is mixed with a hazardous waste, the mixture rule provides that the resulting mixture will be presumed hazardous. The derived from rule states that a solid waste generated from the treatment, storage, or disposal of a hazardous waste (such as treatment sludge, spill residue, ash, or emission control dust) will also be presumed hazardous. If the waste being stored, treated, or disposed of is a *listed* hazardous waste, then the mixture, residue, or derived from waste will also be a listed waste by definition, regardless of its characteristics. Consequently, a broad spectrum of derivative materials, some containing very low concentrations of hazardous constituents, may be regulated under RCRA.

If a material is a hazardous waste, whether listed, characteristic, or hazardous by virtue of the mixture or derived from rule, it is subject to the full range of RCRA regulations. RCRA regulations establish detailed record-keeping, reporting, and transportation requirements for generators and transporters of hazardous waste. The regulations also include comprehensive management standards for *treatment, storage, and disposal* (TSD) facilities including the detailed design, installation, operating and maintenance requirements for different types of hazardous waste management units such as tanks, containers, and surface impoundments as well as training, contingency plan, and financial responsibility requirements.

In addition, a TSD facility must have an RCRA permit unless it qualifies for one of the limited permitting exclusions. For example, short-term storage (90 days) by the generating facility is exempt from permitting, as are totally enclosed treatment units and wastewater treatment tanks.

One of the fundamental problems with the current RCRA regulatory program is that it is an "all or nothing" and "one size fits all" program. Low-risk and high-risk activities and wastes are regulated in the same manner. Consequently, the EPA tends to interpret the definitions of solid waste and hazardous waste very broadly in order to bring as many materials as possible within the scope of RCRA. The EPA fears that any move toward underinclusion could inadvertently exempt potentially hazardous materials and activities from RCRA regulation. However, the result is frequently overregulation in the sense that many low-risk wastes and activities are subject to the full panoply of RCRA requirements.

Recycling. One of the express purposes of RCRA is to encourage "resource conservation and recovery" or recycling. However, RCRA may discourage recycling in some situations. For example, a solid waste is defined as discarded materials which, in turn, are defined to include recycled secondary materials. RCRA regulations specifically exclude certain kinds of recycled secondary materials from regulation, including materials that are used as ingredients in an industrial process to make a product. However secondary materials which are used to make a commercial fuel are *not* exempt and are subject to full RCRA regulation. Furthermore, the EPA has narrowly interpreted the existing recycling exclusions, which means that many reused secondary materials are regulated as hazardous wastes.

The mixture and derived from rules also serve to brand otherwise nonhazardous materials as hazardous wastes, which may effectively preclude recycling. For example, assume a facility uses an incinerator to manage its listed hazardous wastes. Fly ash from the incinerator is captured in an air pollution control device, producing a residue which does not exhibit any of the characteristics of hazardous waste. However, the residue is a listed hazardous waste by virtue of the derived from rule. Recycling facilities would be reluctant to accept this residue because it could trigger RCRA's costly storage and management requirements, as well as permitting and facilitywide corrective action. Further, there may not be a market for the waste-derived product. Consequently, the residue will probably be disposed of in a hazardous waste landfill rather than recycled.

International Agreements

International agreements related to environmental issues are becoming more common and more specific. Environmental issues are now an integral part of international trade discussions, and numerous bilateral and multilateral agreements are devoted specifically to environmental matters. The most prominent of these include the agreements reached at the 1992 Earth Summit in Rio de Janeiro. Perhaps the most important implication of these agreements to the engineering practitioner is the opportunity for technological innovation and technology transfers, particularly as many nations raise their environmental standards and adopt new standards to address global environmental problems.

Trade and the environment

The General Agreement on Tariffs and Trade (GATT) is a free-trade agreement which generally prohibits discrimination against imports. Where protection is given to a domestic industry, GATT contains negotiated tariffs. GATT prohibits nonnegotiated tariffs as well as nontariff barriers to trade such as "green" product standards and certain environmental requirements.

Stringent environmental regulations in a country such as the United States could provide a competitive disadvantage for domestic industries. For example, in adopting the 1990 Clean Air Act Amendments, Congress acknowledged that businesses would incur significant compliance costs, which could impair their competitiveness. Consequently, the statute directs the President to promote harmonization of standards or obtain trade adjustments, as appropriate, to alleviate the competitive disadvantage.

On the other hand, less stringent standards in another country could effectively constitute a subsidy to facilities located in the other country. To counteract this, Senator Boren, a Democrat from Oklahoma, introduced legislation in 1991 which would impose a tariff on imports into the United States, reflecting the costs that a foreign producer would have to incur to meet U.S. environmental standards.

Environmental issues have also been raised in the context of the North American Free Trade Agreement (NAFTA) between the United States, Mexico, and Canada. Opponents of NAFTA argue that increased economic development and traffic along the border with Mexico could lead to increased degradation of the environment. Proponents contend that increased trade in Mexico will result in the availability of more funds for pollution control. Studies have shown, e.g., that once a country reaches a certain gross national product (GNP) per capita, pollution tends to decrease and environmental protection increases.

To alleviate the environmental concerns, a side agreement to NAFTA was negotiated in August 1993 that deals with environmental issues. In essence, the side agreement creates a trinational Commission for Economic Cooperation which can impose trade sanctions for persistent failures to enforce domestic environmental laws.

Multilateral agreements

There are numerous bilateral and multilateral international agreements dealing with environmental issues. As suggested above, the most notable of these were signed at the Earth Summit in Rio de Janeiro in June 1992. The overriding theme of the Earth Summit was "sustainable development," which has become a catchword for the integration of trade and environmental issues. The Rio documents clearly advocate free trade and reflect a conviction that trade liberalization will result in economic growth which, in turn, will provide a vehicle for greater environmental protection.

One of the more tangible agreements from the Earth Summit is the Climate Change Convention, signed by 154 countries. It sets targets for greenhouse gas emissions such as carbon dioxide, which allegedly contribute to global warming. There is considerable debate concerning the causes of global warming or whether global warming is even occurring. Nevertheless, the Climate Change Convention requires parties (including the United States) to reduce greenhouse gas emissions to 1990 levels. President Clinton committed to reach 1990 levels by the year 2000, and in October 1993 he issued his climate change action plan, which was designed to accomplish that goal.

Nearly half of the greenhouse gas emissions in the United States are from the combustion of fuels by electric utilities. Consequently, President Clinton's plan attempts to reduce electricity use, thereby decreasing aggregate energy use by utilities. The plan also encourages the use of cleaner fuels, such as natural gas, rather than coal. Clinton's climate change action plan does *not* direct the EPA to impose emission limitations, as we may have expected in the past. Rather, it consists primarily of voluntary initiatives designed to encourage cost-effective solutions, where possible.

Prior to the Earth Summit, 24 nations signed the Montreal Protocol, which restricts the use of ozone-depleting chemicals such as chlorofluorocarbons (CFCs). The Montreal Protocol is significant in that it sets specific targets and deadlines for phasing out ozone-depleting chemicals. In compliance with the commitments under the Montreal Protocol, Congress adopted implementing legislation as a part of the 1990 Clean Air Act Amendments, and the EPA is in the process of issuing corresponding regulations.

Implications for Pollution Prevention

The engineering practitioner should keep in mind that environmental regulations could have a significant impact on the design and feasibility of a project. For example, if a project requires an air permit, the corresponding delay and administrative costs could mean that the project is no longer financially feasible. In addition, a unit which would be considered a process unit from an engineering standpoint could be considered a waste treatment unit from a regulatory standpoint, which could trigger full RCRA regulation. Further, regulations or permits may specify required technologies or may discourage innovative technologies. The current regulatory focus on end-of-pipe controls, rather than pollution prevention, can often discourage otherwise technically desirable projects.

Nevertheless, there is an emerging focus on and commitment to pollution prevention, particularly by the EPA. On June 15, 1993, EPA administrator Carol Browner issued a policy statement which purports to make pollution prevention an integral part of EPA activities. As stated in the policy statement:

> When EPA was created in the early 1970's, our work had to focus first on controlling and cleaning up the most immediate problems. Those efforts have yielded major reductions in pollution in which we should all take pride. Over time, however, we have learned that traditional "end-of-pipe" approaches not only can be expensive and less than fully effective, but sometimes transfer pollution from one medium to another. Additional improvements to environmental quality will require us to move "upstream" to prevent pollution from occurring in the first place.

In reaching this objective, the EPA intends to integrate pollution prevention concepts into its regulations and compliance programs. For example, the EPA is evaluating a cluster approach to regulations, which would involve adopting comprehensive, multimedia regulations tailored to a particular industry. The agency also intends to employ more voluntary programs such as those used in Clinton's climate change action plan, designed to take advantage of site-specific opportunities for reducing pollution.

3

Waste Minimization: Incentives and Barriers

Richard J. Camm
President
Linnhoff March, Inc.
Houston, Texas

Peter M. Nuttall, Ph.D.
Independent Environmental Consultant
Crewe, England

Introduction

The traditional approach to waste management has tended to focus on end-of-pipe technologies, since these were the first available. Such "sink it," "burn it," or "bury it" solutions have come under increasing scrutiny, as discussed in the preceding chapters. Recently, many individuals and organizations have been promoting the concept of waste management on which the preferred option is waste prevention, either through product substitution or process replacement, and source reduction through equipment redesign, product formulate or process modification and improvement.

The shift in emphasis, then, is from treatment and disposal to waste minimization through prevention and reduction. It is this area that is receiving increasing support, not only from regulators and the community, but also from the industrial waste producers themselves. There are increasing numbers of incentives which are gradually moving the policies and technological acceptance forward. However, there are also several barriers, both economic and attitudinal, which are inhibiting the rapid introduction of waste minimization policies into industry. These are discussed below with reference primarily to the situation in Europe. The chapter concludes with several case studies that illustrate how the barriers are being overcome and waste minimization concepts and technologies are being implemented in industry.

Incentives

There are economic, government, and corporate incentives for industrial waste minimization. Moreover, in a broader perspective, there are global considerations that provide further major incentives for reducing emissions. These issues are discussed below.

Economic incentives

There are powerful economic incentives for a company to establish a waste minimization program. Strong among these is the improved profitability which can be gained through undertaking waste minimization at the source by increasing operating efficiency and improving process reliability. Liquid, solid, or gaseous waste materials are inevitably generated during the manufacture of any product. Apart from causing potential environmental problems, these wastes represent process inefficiencies in the production process leading to losses of valuable raw materials and energy. The wastes also require significant investment in pollution control. Waste minimization offers the most cost-effective alternative to treatment and disposal.

Most end-of-pipe treatment or disposal options really transfer the problem elsewhere. There are many additional (economic) penalties of this approach:

- The cost of waste handling, treatment, and disposal, together with on-site waste monitoring and control, can represent a significant financial commitment by the company. For instance, in the U.S. refining industry alone, the investment costs on air pollution control and on wastewater treatment are estimated at $6.8 billion and $3.02 billion, respectively, for the period from 1991 to 1995. The corresponding operating costs are put at $1.9 billion and $0.9 billion. The operating costs are expected to increase by 50 to 100 percent over 10 years.[1]

- Many countries have a requirement by government environmental agencies for the company itself to undertake routine self-monitoring of waste streams for compliance, and this can mean additional costs in analytical charges, staff time, and training to the company. Waste generation costs will, by definition, include administration and paperwork at each stage in the waste production and disposal pathway. In Europe, Australia, and the United States, a manifest system for special wastes includes a consignment note register to be kept by producers, carriers, and disposers, adding to the administrative costs of waste transfer. These costs must eventually pass back to the company which generated the waste product.

- On-site storage space may be required for waste storage, which is expensive and otherwise could have been used for productive operation. With on-site storage, there is a legal obligation to undertake correct containment and waste control, because if leakage or spills occur, then the company operators will be made liable. The company may be asked for an environmental bond as a guarantee to finance the cleanup of a pollution spill, particularly where there is a history of significant waste generation released into the environment or where the industry is located in an ecologically sensitive area.

Government incentives

Lack of capital to fund the waste minimization program and lack of expertise can severely hamper a company's efforts to identify and implement waste reduction projects. In certain cases, financial assistance and technical support can be obtained from government agencies or through government-sponsored programs, e.g., from the Department of Trade and Industry (DTI) in the United Kingdom, European funding schemes (European Community), and the Department of Labour and Industry (DACTI) in Australia. There are also numerous federal, state, and local government programs that promote pollution prevention in the United States. Chapter 21 describes an example of this—the Amoco–U.S. EPA Pollution Prevention Project at Amoco's Yorktown, Virginia, refinery. In this project, both parties contributed both financially and technically in a unique voluntary study of pollution prevention opportunities at an industrial facility.

There are a great many government incentives in the United Kingdom (UK) and Europe. A selection of these is described below:

UK government incentives. In 1990 the UK government set up a grant aid program to promote innovation in environmental technology. The program included existing ETIS and DEMOS schemes after their 3-year trial. The ETIS scheme provided grant aid for the development of cleaner manufacturing processes and pollution abatement techniques, while DEMOS was aimed at promoting the spread of good environmental management practices and adoption of clean technologies. Both schemes fell short of expectations. Only £8.5 million of the £16 million allocated to them was used. They also proved extremely demanding to manage with decisions on grant application alone taking an average of 12 months. This program was dissolved in late 1993.

The grant-in-aid program is to be replaced by a £16 million scheme to encourage business to adopt established methods of reducing waste and pollution at the source. The new environmental technology best-practice program is to be run jointly by the Department of Environment (DoE) and the Department of Trade and Industry (DTI). The inspiration for the plan was generated in part by the Aire and Calder Project in Yorkshire which confirmed that many financially attractive waste minimization opportunities are waiting to be seized by applying existing technology and good management practice.

Best-practice program. This program has been operating since 1989 with an overall goal to generate energy savings of £800 million per year by the year 2000 in industry and commerce. In the process the intention is to cut UK carbon dioxide emissions by 5 million tons (as carbon) per year by that date. It works primarily by disseminating information via seminars, good-practice guides, and case studies.

Environmental Enquiry Point. In the United Kingdom, the Department of Trade and Industry has an environmental Enquiry Point, which is a telephone inquiry service for UK companies only, run by Warren Spring Laboratory (toll-free 0800 585 794). The service provides information and advice on a wide range of environmental issues affecting business, including technical matters, existing legal requirements, new standards, and forthcoming conferences and seminars. Warren Spring Laboratory has recently been merged with AEA Technology to form the National Environmental Technology Centre.

The DTI Enterprise Initiative. Consultancy initiatives under the DTI Enterprise Initiative provide small and medium-size UK firms access to assisted consultancy in the areas of marketing, quality, design, manufacturing systems, business planning, and financial and information systems (toll-free 0800 585 794). These can be used to obtain independent expert advice on a range of environmental problems related to product design and manufacture, waste minimization, and other aspects of waste management.

The teaching company scheme. This scheme (0367 22822) provides on-the-job support from higher education institutions and graduates to companies intending to introduce major operational changes.

Environmental protection technology (EPT) scheme. The EPT scheme is a Department of Environment UK initiative to encourage technical innovation in the field of environmental protection. The EPT unit is located at Warren Spring Laboratory (0438 741 122). R&D grants are available in a number of priority areas.

Collaborative research and development. The DTI offers grants for collaborative research under a number of schemes which could help in the development of new technologies to tackle environmental and waste management problems.

European Community (EC) incentives

The EC LIFE program. The LIFE program is the main source of EC funding for environmental initiatives throughout Europe. The general objective of LIFE is to contribute to the development and implementation of the EC environmental policy and legislation. LIFE is an EC instrument reserved for the environment which has been allocated ECU 440 million until the end of 1995. The scope of LIFE is broadly defined by the promotion of sustainable development and quality of the environment, protection of habitats, environmental services, education and training, and actions outside the EC territory.

LIFE's priority is to establish and develop new clean technologies which create little or no pollution and make few demands on resources; and to devise and develop new techniques for the collection, storage, and recycling of wastes and locating and restoring contaminated sites.

THERMIE program. The THERMIE program is a funding agency which promotes European energy technology and is a vital part of the EC strategy for meeting energy challenges, particularly through reducing emissions of greenhouse gases.

CRAFT. CRAFT is a funding agency which is an EC initiative that facilitates participation of small and medium-size industrial enterprises in research. Key areas of research include waste reduction, recovery, recycling, and reuse of industrial wastes.

Corporate incentives

Application of waste minimization demands the provision of useful alternatives to waste treatment and disposal through the reduction of waste at its source. It is not a panacea for zero waste and should be introduced only in situations where the company has a commitment to changing the traditional approach. Two specific incentives for making this commitment are outlined below:

Corporate image. The corporate image is inevitably affected adversely by repeated infringements on compliance. Conversely, if a company can demonstrate that it is going beyond legislative requirements and is attempting to improve its environmental performance, this will create a positive public image. This, in turn, can be utilized as a marketing tool: A public demonstration of improved environmental performance can improve company performance within the marketplace, which is increasingly sensitive to environmental issues. Improved environmental performance also enhances the company's image in the eyes of its shareholders, employees, and community.

Technical reasons. Failure to address environmental issues adequately can create an occupational hazard for the workforce and can affect the health of the entire community. Consequently, there are compelling technical and health-related reasons for treating waste streams and/or minimizing wastes and emissions.

Global concerns

There is increasing sensitivity in the international community to environmental concerns which implicate industrial emissions such as global warming, land care, air quality, public health, biodiversity, acid rain, and sustainable development. These concerns provide an additional strong incentive for industrial waste minimization.

Greenhouse gas emissions. The greenhouse effect is a natural phenomenon that keeps the earth's surface within a temperature range necessary to sustain the current diversity and distribution of life through the existence of levels of certain gases in the atmosphere. In recent times, human activities have caused these levels to increase, which will result in an enhanced greenhouse or warming effect. Increasingly, the global community is being alerted to the potential for changes to the earth's climate as a result of rising levels of greenhouse gases in the atmosphere.

The Intergovernmental Panel on Climate Change (IPCC), established jointly in 1988 by the World Meteorological Organization and the United Nations Environment Program (UNEP), released its scientific assessment of the greenhouse effect in July 1990,[2] updated in 1992.[3] These reports essentially noted that emissions resulting from human activities are substantially increasing concentrations of carbon dioxide (CO_2), methane (CH_4), chlorofluorocarbons (CFCs), and nitrous oxide (N_2O). The global mean surface air temperature has increased by 0.3 to 0.6°C in the last 100 years. Should greenhouse gas emissions continue to increase at the current projected rate, an average rate of global mean sea-level rise of 3 to 10 cm per decade is predicted over the next century, resulting mainly from thermal expansion of the oceans.

Currently, all industrial nations have agreed to an interim planning target to limit greenhouse gas emissions (carbon dioxide, methane, nitrous oxide) not controlled by the Montreal Protocol on ozone-depleting substances. This target seeks to achieve 1988 levels by the year 2000 and to reduce these emissions by 20 percent by the year 2005. A number of countries have stated that they would not proceed with the adoption of subsequent response measures since, in the absence of similar action by major greenhouse gas-producing countries, there would be net adverse economic impacts nationally or on trade competitiveness, without significant environmental benefits.

Carbon tax. The European Economic Community has adopted carbon energy tax plans, although the United Kingdom abstained. The official view is that the EC goal of CO_2 emissions at 1990 levels by the year 2000 can be met. Such an approach has not yet reached the United States, although there is a general belief that it will at some future time.

Ecological sustainable development (ESD). The proposed goal of ESD is development that improves the quality of life, both now and in the future, in a way that maintains the ecological processes on which life depends. The objectives of ESD are to enhance individual and community well-being and welfare by following a path of economic development that safeguards the welfare of future generations, to provide for equity within and between generations, and to protect biological diversity and maintain ecological processes and systems.

Agriculture, forest use, fisheries, manufacturing, mining, urban and transport planning, tourism, and energy use and production form the sectoral issues for discussion. Environmental protection through waste control and reduction emerges as a critical component of sustainable development.

Barriers to Waste Minimization

There are many barriers to industrial waste minimization. The costs of implementation can be a significant economic deterrent. Regulations that are intended to restrict emissions can have the effect of discouraging innovative waste minimization options. Waste minimization technology is still considered high-risk. Corporate cultures are also often unreceptive to waste minimization concepts. These issues are addressed below.

Economic barriers

Any waste minimization program will require capital for initiation and full establishment. Although reducing waste will bring down operating costs and improve revenue (see, e.g., Chap. 22) and the overall environmental position of the company will be enhanced, it is often difficult to quantify the value of these items against a "hard" economic need such as capital. Any demand for capital has to take its turn in overall assessment of value.

The true costs of waste generation need to be correctly identified. A total cost accounting system is needed which will avoid the situation which must occur repeatedly, in which environmental costs are assigned to overheads or accepted as a recoverable cost of production. Companies need to track and identify all financial costs of pollution prevention and to properly account for benefits.

Resistance to waste minimization may also be offered if there are any indications at all of increased production costs, increased raw material costs, or the need for new or modified equipment. Any economic evaluation of costs and benefits for a waste minimization strategy will initially consider reliable, verifiable, and tangible information. This can result in the strategy being abandoned because of unattractive economics. It is essential that proper account be taken of less tangible benefits, such as those mentioned above, and prospective legislation and taxes. In this way, the true worth of waste minimization is assessed, and projects will begin to offer attractive benefits.

Regulatory barriers

The rapid growth of industrial products and technology over the years since the first industrial revolution has led to a similar rapid growth in environmental discharges and pollution. The state of the environment prompted reform as governments became aware of the consequences of uncontrolled releases and dumping. In England the first laws were introduced to protect the environment in the Gas Works Clauses Act of 1847, which prohibited the discharge of gas waste into streams.

Today there are numerous bills and pieces of legislation which constrain noise, odor, air emissions, solid waste transport and disposal, and wastewater discharges beyond the boundary of the plant, together with a multiplicity of laws and regulations dealing with health and safety, on-site contamination, work practices, efficient technology, and upgrading within the premises.

Legislation has to keep abreast of industrial advancement and environmental protection; otherwise, we could still be attempting to use sewage farms to treat polychlorinated biphenyls. The improvements in environment are achieved by increasing emission constraints, since this is the most direct approach which can be regulated and enforced. Consequently, as the environmental constraints tighten, treatment and disposal become more and more costly and the option for waste minimization becomes increasingly attractive.

There are trends in statutory controls that should facilitate the ability to undertake waste minimization. However, much of the existing legislative framework actually presents significant barriers to waste minimization. Often a process change requires a reapplication for an environmental license which takes many weeks to obtain. Reapplications often have to be supported by payment, the industrial process comes under close scrutiny from the government agency, and government employees often demonstrate only a passing familiarity with the technology and need explanations throughout the process, all of which takes time, commitment, and trust from the company. Frequently, a company investing in innovative waste minimization attracts more state attention than necessary, with more government departments becoming involved in a learning curve than are strictly warranted. External facilitators are often needed to lead the company through the environmental legislation maze. Company public relations and good communication can normally clarify any problems which might be perceived by government, but the whole procedure is typically time-consuming and costly.

The concept of criminal liability in the United States may impede the ability of a company to move away from the compliance approach into what is believed to be the less charted waters of waste minimization.

One other key point with respect to regulatory barriers and environmental legislation is technological: Regulatory constraints can be *quantified,* and end-of-pipe treatment technology is sufficiently well established to give good expectation of compliance. Waste minimization is not perceived in this light. This is discussed in more detail under "Technological barriers."

Technological barriers

Legislative constraints are generally quantified and represent "hard targets" which must be achieved. Failure to attain these targets can lead to severe penalties.

End-of-pipe treatment is well established and predictable and is regarded as reliable to achieve a particular objective. For instance, meeting a particular maximum for HCl concentration in gas emissions or a maximum annual quantity of CO_2 emitted is relatively simple. Scrubbing or absorption is well understood and quite predictable. This approach may not offer the best potential benefits, but it does offer the *lowest risk* to management.

On the other hand, most waste minimization procedures are in their infancy. Although good results are achieved, e.g., by modifying a process, the final emission levels are not predicted with great confidence. What would be the reduced HCl concentration due to the redesign of early stages in the process? This typical objection will diminish over time as technology matures and the precedents of a large number of case studies can be followed.

There does need to be a stronger emphasis from technological developers on predictability and the ability to satisfy the concerns of management with respect to compliance.

Corporate barriers

Corporate barriers represent the resistance within organizations to proceed with waste minimization programs for a wide range of reasons, most of which can be classified as either cultural or technical.

Cultural barriers

- Lack of senior management commitment
- Poor internal communications
- Lack of awareness
- Restrictive employment practices
- Bureaucracy
- Plant manager efficiency measured on production figures, not waste minimization objectives
- Waste minimization not regarded as cost-effective

Cultural barriers to utilization of waste minimization procedures can be overcome with training programs to encourage all personnel at all levels to participate in waste minimization, incentive schemes, bonus or shares of the saving made by waste minimization, lump-sum awards, company newsletter recognition, etc. However, the most important consideration remains a genuine commitment of senior management to this activity. A very successful program initiated by Dow U.S.A. is described in Chap. 22.

Technical barriers

- Lack of suitable information on techniques for waste minimization
- Source reduction leading to noncompliance discharge or emission
- Concerns about changes to product quality and customer acceptance
- Changes to incorporate waste minimization causing shutdown of existing operation for an unacceptable period
- Performance of new operation not as expected
- Lack of space for additional equipment
- Poor reaction from employees

To counter these obstacles, the company and its employees need to be alerted to new and updated information from the government, trade associations, professional institutes, consultants, externally published articles, and in-house newsletters. Customer concerns can be addressed by carefully assessing the customer's needs, implementing a pilot trial of new processes and products, and increased quality control.

Facility operational downtime can be reduced by involving design and production personnel in the planning phase.

To conclude, the above barriers represent a formidable challenge to overcome. The solutions will come with continual education and training of staff on new approaches, in new technology, and in many case studies. The introduction of incentives from government and other outside sources, as discussed earlier in this chapter, will be major catalysts in the process.

Case Studies

Chapters 21 through 23 describe a variety of U.S. waste minimization programs in some detail. These include programs sponsored by government agencies, utilities, and industry. In this chapter, four short European case studies are presented to illustrate how the incentives for waste minimization provide a driving force for process improvement, and how the barriers to waste minimization can be and have been overcome.

Landskrona study

One of the first waste minimization projects was the Landskrona waste minimization study, which was started in October 1987.[4] The aim of the project was to explore economic and environmental benefits of liquid waste and air emission reductions from seven small companies comprising graphics, metalworking, and chemical industries located in the Landskrona region of Sweden. The study was funded by the Foundation of REFORSK, the National Swedish Board for Technical Development (STU), the National Swedish Industry Board (SIND), and local authorities of Landskrona, with complementary research undertaken by the Environmental Research Unit (TEM) at the University of Lund.

The project successfully identified cost-effective multimedia waste minimization options for companies in a number of industrial sectors, including printing, wheel and brake manufacture, light fixture manufacture, electroplating, and chemical production. Corporate and government interest in the project stimulated support for other projects throughout Swedish industry. However, company directors were initially skeptical about the benefits of undertaking a waste minimization study, and TEM found that this misconception was a primary obstacle to project establishment.

PRISMA study

A similar approach to Landskrona was adopted in Holland in late 1988.[5] The PRISMA study involved 10 companies from five industrial sectors: food processing, metal finishing, metalworking, public transport, and chemical manufacturing. The scheme set out to show that pollution prevention in terms of concept, management, and technology could be implemented successfully. It proved that relatively small changes in industrial housekeeping and technical modification had a significant impact on reducing waste at a relatively low cost.

The study which was funded by the Netherlands Technology Assessment Bureau (NOTA) and Ministry of Economic Affairs, with input from Universities of Amsterdam and Rotterdam, was prompted by a U.S. EPA manual.[6] A total of 164 waste minimization options were identified in the 10 companies, of which 111 were considered feasible. Of these, 45 were implemented during the project, which lasted 2 years. Out of this group of projects, 80 percent either cost nothing or had a payback period of less than 1 year.

One participant (DSM Resins) made major savings on two of its waste streams. One of its production processes used an earth filter on a perforated plate. This retained a significant amount of resin product and required washing between batches, which consumed 38 m^3 of toluene or xylene and 108 m^3 of caustic rinse annually. By replacing the earth filters with candle filters, DSM drastically reduced the consumption of rinse liquids and minimized product loss. The new filters paid for themselves in 1 year.

One of the key conclusions of the study was that waste reduction options can be realized in short time scales. In addition, the companies found that there was an overall improvement in the quality of existing products as well as an increase in productivity. Nevertheless, the study noted that small companies committed to waste reduction often do not have the financial resources or technical skills to pursue their aims.

PREPARE

Because of the success of PRISMA, the Dutch government in mid-1990 set up a working group of representatives from other European governments and industry under the title Preventive Environmental Protection Approaches in Europe (PREPARE). This group undertook the preparation and provision of information on waste minimization for industry and identified the need for new projects elsewhere in Europe. The working group has been successful in promoting a number of projects on waste minimization throughout Europe, including Austria, Denmark, Finland, and Belgium. Significantly, PREPARE is the first international exchange and cooperation scheme in the arena of clean technologies.

Bayer Leverkusen

Our final example is a systematic study of CO_2 emission reduction options undertaken by German chemicals giant Bayer at its Leverkusen factory. The study was based on total site pinch analysis (see Chap. 7) and was carried out by Bayer as part of an effort to reduce combustion-related emissions. The German Federal Government has set a goal of achieving a 25 percent reduction in CO_2 emissions (relative to a 1987 baseline) by the year 2005. This, together with a recognition of the high cost of energy (10 to 20 percent of production costs) and the adverse ecological effects of emissions, provided the motivation for this work.

Leverkusen is Bayer's largest facility, with a wide range of chemical processes—batch and continuous, organic and inorganic—on the site. The factory's power station burns coal and fuel gas to produce steam at 110 bar. This is let down through steam turbines to 31 and 6 bar and is used at these levels for process heating.

The study identified a large number of opportunities for reducing CO_2 emissions and at the same time lowering energy costs. These included heat integration and various process changes within the individual production units as well as modifications to the steam/power system. Some of these projects are more expensive to implement than others, and the extent to which CO_2 emissions can be reduced is a function of the payback period. The maximum theoretical scope for CO_2 reduction was shown to be 28 percent. However, if only projects with an incremental payback period of less than 3 years are implemented, this goes down to 8 percent.[7]

Conclusion

Waste minimization by means of process modification and process design potentially offers a very powerful procedure to reduce pollution, waste, and emissions and at the same time to increase efficiency and reduce operating cost. The procedures involved are not well developed, and some emphasis is required in certain areas to promote the regular use of waste minimization programs, including

- Technological improvement
- Greater corporate awareness, particularly regarding case studies
- Government incentives
- A move to economic evaluation of waste generation and waste minimization, particularly appropriate accounting practices
- Support from regulatory authorities

The case for replacement of end-of-pipe solutions by waste management and minimization strategies is a strong one, and industry is to be expected to move fully this way over the next few years.

References

1. J. G. Musgrove and E. J. Swain, "U.S. Refinery Compliance Costs—The Elements of an Estimate," 86th Annual Meeting of the Air and Waste Management Association, Denver, June 13–18, 1993.
2. InterGovernmental Panel on Climate Change (IPCC) Working Group I (J. T. Houghton, G. I. Jenkins, and J. J. Ephraums, eds.), *Climate Changes: The IPCC Scientific Assessment,* Cambridge University Press, Cambridge, UK, 1990.
3. InterGovernmental Panel on Climate Change (IPCC) Working Group I (J. T. Houghton, G. I. Jenkins, and J. J. Ephraums, eds.), *Climate Changes: The IPCC Scientific Assessment,* 1992 IPCC Supplement, World Meteorological Association, United Nations Environmental Program, Cambridge University Press, Cambridge, UK, 1992.
4. M. Backman, D. Huisingh, E. Persoon, and L. Siljebratt, *Preventative Environmental Protection Strategy: First Results of an Experiment in Landskrona, Sweden,* TEM/University of Lund, Sjobo, Sweden, August 1989.
5. R. van Berkel, M. Crul, S. De Hoo, and P. Koppert, *Business Examples with Waste Prevention: Ten Case Studies from the Dutch PRISMA Project,* Dutch Ministry of Economic Affairs, Leiden (ISBN 90 346 2565 6), 1992.
6. Environmental Protection Agency, *Waste Minimization Opportunity Assessment Manual,* EPA Hazardous Waste Engineering Research Laboratory, Office of Research and Development, Cincinnati, Ohio, 1988.
7. M. Bueb and J. Kussi, CO_2-*Reduzierung durch Energieeinsparung in der chemischen Industrie,* GVC-Jahrestagung 1993, Nürnberg, September/October 1993.

Pollution Prevention and Process Integration— Two Complementary Philosophies

Alan P. Rossiter, Ph.D., and Jimmy D. Kumana
Linnhoff March, Inc.
Houston, Texas

Introduction

Historically, pollution has been considered an unavoidable by-product of industrial progress. However, as the preceding chapters have shown, perceptions have changed. It is no longer socially, legally, or politically acceptable to pollute, and industry has to respond by developing "greener" processes. Rather than generating and discharging wastes, we must find ways either to make products without creating pollution or else to recover and reuse the materials that we have hitherto considered "wastes." This philosophy has come to be known as *pollution prevention*.

Concurrently with the changes occurring in perceptions of pollution, fundamental changes have been taking place in the way in which process engineering activities are carried out. At the heart of this is the development of the field of process integration. This discipline encompasses a number of interrelated methodologies for designing and revamping industrial processes, based on a philosophy of process design that recognizes the unity of the whole process.

Process integration is highly compatible with the philosophy of pollution prevention, and complementary to it. One of the main characteristics of process integration techniques is that they are inherently conservation-oriented, i.e., they enhance process efficiency by minimizing the use and/or maximizing the recovery of energy and materials. This links them directly to the goals of pollution prevention. Moreover, pollution prevention depends on *overall* process designs that are intrinsically environment-friendly, rather than simply adding pollution control equipment. As process integration techniques provide a basis for analyzing and developing designs in their entirety, they can readily be focused on pollution prevention objectives.

Pollution Prevention and Its Hierarchy

Until recently, efforts to reduce environmental emissions have emphasized pollution *control*, i.e., minimization of the harmful impact of toxic substances before their discharge to the environment. Typically this involves providing end-of-pipe solutions to environmental problems, e.g., scrubbing or incinerating to remove toxic vapors from exhaust airstreams or neutralizing acidic or alkaline effluents.

Unfortunately, the pollution control approach has a number of shortcomings, as discussed in Chap. 1. It provides the basis for most of the existing environmental regulations, which impose a disproportionately large financial burden on process operators. Moreover, end-of-pipe control techniques typically offer no real reduction in pollution; rather, they simply transfer pollution problems from one medium to another. For example, absorption of undesirable vapors may mitigate an air emission problem, but at the expense of creating a water discharge problem. Sometimes the same problem appears in reverse, e.g., when air stripping is used to remove volatile contaminants from water. Other techniques simply disguise real problems: It is a common "pollution control" practice for operators of processes with gaseous and aqueous emissions to dilute the discharges to comply with mandated concentration limitations. "Dilution is the solution to pollution!"

There is now an increasing philosophical shift away from pollution control toward pollution *prevention*. The difference is basically this: Pollution control aims to prevent the release of undesirable materials and relies heavily on end-of-pipe technologies to achieve this objective. Pollution prevention seeks to eliminate pollution problems by not creating potential pollutants at all or, if their creation cannot be avoided, by not letting them reach the end of the pipe if this can reasonably be avoided. However, it is not always economically (or even technically) feasible to eliminate the production of process wastes completely. Consequently, a certain amount of end-of-pipe control is often required, even within the pollution prevention framework.

The pollution prevention hierarchy[1] ranks the basic means of reducing emissions in terms of global desirability. Several versions of the hierarchy have been proposed. One of these (showing the ranking from most to least desirable) is given below:

1. *Source reduction,* e.g., use cleaner feed material and eliminate leaks.

2. *Reuse and recycle,* e.g., reuse waste from one unit as a feed to another, or recycle contaminated material to the unit that generated it, with a small purge to maintain acceptable contaminant levels.

3. *Treatment,* e.g., acid neutralization or incineration.

4. *Safe disposal,* e.g., landfills. Disposal should be minimized, in terms of both quantity and potential harmfulness of the material discharged.

Any pollution prevention option can, in principle, be classified under one of these categories, although in practice the classifications are not always clear-cut. For example, treatment options could be used to make a stream suitable for reuse or recycle (these are commonly known as *regeneration* options); or they could be precursors to safe disposal (end-of-pipe treatment).

Pollution prevention has several clear advantages over pollution control. Any pollutant that reaches the end of the pipe represents a loss of raw material. Since pollution prevention seeks to minimize the amount of material that suffers this fate, it is a good way of conserving natural resources, which is good stewardship. This also results in savings in raw material costs for the process operator and thus tends to improve process economics.

There are also some difficulties inherent in the pollution prevention concept. From the perspective of the regulators, it is far more problematic to legislate and monitor than conventional pollution control. Pollution prevention is not usually accomplished simply by placing a piece of equipment immediately ahead of a discharge point. Identifying good solutions often requires a high degree of technical expertise and creativity, and ideas are not always directly transferable, even between outwardly similar industrial facilities.

Pollution prevention poses a unique challenge for those engaged in the design and improvement of industrial processes, and this challenge forms the heart of this book. As already noted, pollution prevention cannot be accomplished simply by placing a piece of equipment immediately ahead of a discharge point. Rather, it depends on overall process designs that are intrinsically environmentally friendly. This calls for a philosophy of process design that recognizes the unity of the whole process: process integration.

Process Integration Methods

A number of different systematic approaches to industrial process design have been developed under the banner of process integration, and they are fundamentally changing the way in which process design and retrofit activities are carried out. These methodologies do not, in general, attempt to invent new types of equipment or unit operations. Rather, they focus on ensuring that existing process technologies are selected and interconnected in the most effective ways, e.g., constructing heat exchanger networks with the optimum balance of capital and energy costs.

Typically the procedures start with an overview of the process as a whole, rather than focusing on individual unit operations or pieces of equipment. In this way a "correct" structure can be developed for the overall plant, with individual items of equipment being fit into this structure. This approach has been successfully applied for many years in cost minimization activities, with the emphasis on the trade-off between capital and operating costs. However, the same basic methods can also be used to explore the three-way trade-off between capital cost, operating cost, and environmental impact.

Process integration methodologies can be classified under three major headings.[2] Each of the following three parts of this book is devoted to one of these categories and its uses in pollution prevention. The classifications are listed below, with the corresponding part and chapter numbers:

- Pinch analysis: Part 2 (Chaps. 5 through 8)

- Knowledge-based approaches: Part 3 (Chaps. 9 through 14)

- Numerical optimization approaches: Part 4 (Chaps. 15 through 18)

The first chapter in each part introduces the principles of the methodologies within the relevant category. Each of the subsequent chapters focuses on one particular methodology and provides examples of specific applications.

A brief description of each of the categories is given below.

Pinch analysis

Pinch analysis[3] (or *pinch technology,* as it used to be known) is a systematic technique for analyzing heat flows through industrial processes, based on fundamental thermodynamics. It is widely used to determine the scope of energy savings in industrial operations and to define possible process changes to reduce intrinsic energy consumption. The trade-off between energy consumption and capital investment can be assessed.

In the environmental context, this approach is useful primarily as a means of determining the extent to which energy consumption can be reduced, with attendant reductions in NO_x, SO_x, and CO_2 levels.[4] However, it also provides some guidance in the generation of design options for the reduction of other, process-related emissions.[5,6] The pinch method has also been adapted for analyzing mass transfer, especially in dilute aqueous systems,[7] and this offers a new approach to the analysis of wastewater systems.

Knowledge-based approaches

The term *knowledge-based system* (or *expert system*) is used to describe a class of artificial intelligence applications that embodies a system of rules based upon an area of expert knowledge. In this discussion, the term *knowledge-based approaches* is used in a broader sense, describing all process synthesis and process integration methods that build upon an accumulated knowledge base of proven ideas. In the context of waste minimization, this includes methods in which specific pollution prevention ideas are transferred directly from one application to others;[8,9] hierarchical design and review procedures, in which the logical sequence of flow sheet evolution provides a framework for identifying and evaluating waste minimization options;[10,11] and, of course, artificial intelligence,[12] where computer programs mimic human thought processes to develop "clean" process designs.

These approaches can be used either to develop new designs or to identify retrofit options, often starting with minimal data. In new plant design, use of this type of procedure generally results in one or more good designs for the process (i.e., cost-effective, with low emissions). In retrofits, they typically generate a list of potential process improvements for use in revamp projects.

Numerical optimization approaches

There are a variety of numerical optimization approaches, ranging from simulation using simplified mathematical models of the process to sophisticated mathematical programming methods. These are often combined with cost equations to quantify the impact of design decisions on process economics. Graphical representations or cost diagrams can be generated to provide a visual representation of the effect of varying design and/or operating parameters.[13]

These approaches have been applied to a number of environmental problems. For example, simple mathematical models have been used to develop cost versus emissions limit curves.[14] These enable engineers and regulators to explore the impact of process changes on both cost and emission levels, and thus to define the most cost-effective means of achieving any given emissions target. More sophisticated techniques (linear and nonlinear programming) have been used in many different applications. These include minimization of water use and wastewater generation rates from production facilities,[15] waste minimization in pulp and paper operations,[16] and synthesis of reverse-osmosis networks for waste minimization.[17] Similar methods have also been used for real-time optimization to minimize emissions; e.g., mixed integer linear programming (MILP) techniques can be used for optimizing the selection of on-line plant utility system equipment to minimize NO_x emissions.[18]

Application and Implementation

The three different branches of process integration have somewhat different areas of application and tend to yield different, complementary results. Consequently, several different methods are often used together to address a given design problem. The numerical modeling approach is best suited to situations where only a limited number of well-defined design options require evaluation. For complex processes with multiple variants, the simulation effort can become overwhelming, and the other approaches—especially knowledge-based methods—are needed to identify potentially attractive options and narrow down the scope of the problem. The pinch approach is good for identifying fundamental insights into heat-transfer and mass-transfer problems, which can result in step-change design improvements. These differences and complementary characteristics are illustrated in the chapters which follow.

Achieving a successful outcome from waste minimization studies requires more than good technical approaches, however. The technology needs to be managed appropriately, and an environment is needed in which the challenge of waste minimization can be met with a creative, innovative attitude. The final part of the book (Part 5, Chaps. 19 through 23) addresses these issues and describes methods for managing the technology and programs that have succeeded in identifying waste minimization opportunities.

Conclusions

To prevent pollution, it is necessary to adopt a philosophically complementary approach to process design. Process integration provides that approach.

Under the umbrella of process integration, there are a number of systematic design methods that have been successfully used in pollution prevention applications. These can be classified under three main headings—pinch analysis, knowledge-based approaches, and numerical optimization approaches. Each of the following three parts of this book is devoted to one of these classes of process integration techniques and its uses in pollution prevention.

References

1. H. M. Freeman, *Hazardous Waste Minimization,* McGraw-Hill, New York, 1990, p. 7.
2. T. Gunderson and L. Naess, "The Synthesis of Cost Optimal Heat Exchanger Networks," XVIII European Federation of Chemical Engineering Congress, Giardini Naxos, Italy, April 26–30, 1987.
3. B. Linnhoff, D. W. Townsend, D. Boland, G. F. Hewitt, B. E. A. Thomas, A. R. Guy, and R. H. Marsland, "User Guide on Process Integration for the Efficient Use of Energy," *IChemE,* Rugby, England, 1982.
4. R. D. Sung, J. D. Kumana, and A. P. Rossiter, "Reducing Air Emissions by Process Heat Integration," American Society of Mechanical Engineers EcoWorld Conference, Washington, D.C., June 1992.
5. H. D. Spriggs, R. Smith, and E. A. Petela, "Pinch Technology: Evaluate the Energy/Environmental Economic Trade-Offs in Industrial Processes," conference on *Energy and the Environment in the Twenty-First Century,* Cambridge, MA, March 1990.
6. A. P. Rossiter, M. A. Rutkowski, and A. S. McMullan, "Pinch Technology Identifies Process Improvements," *Hydroc. Proc.,* 70(1):63–66, January 1991.
7. Y. P. Wang and R. Smith, "Wastewater Minimization," *Chem. Eng. Sci.,* 49:981–1006, 1994.
8. K. E. Nelson, "Use These Ideas to Cut Waste," *Hydroc. Proc.,* 69(3):93–98, March 1990.
9. S. A. Slater, J. Khan, and R. L. Smith, "Transferring Waste Minimization Solutions between Industrial Categories with a Unit Operations Approach: I. Chemical and Plating Industries," American Chemical Society I&EC Special Symposium on Emerging Technologies for Hazardous Waste Management, Atlanta, September 21–23, 1992.
10. J. M. Douglas, "Process Synthesis for Waste Minimization," *Ind. Eng. Chem. Res.,* 31:238–243, 1992.
11. A. P. Rossiter, H. D. Spriggs, and H. Klee, Jr., "Apply Process Integration to Waste Minimization," *Chemical Engineering Progress,* 89(1):30–36, January 1993.
12. T. F. Edgar and Y. L. Huang, "An Artificial Intelligence Approach to the Design of a Process for Waste Minimization," American Chemical Society I&EC Special Symposium on Emerging Technologies for Hazardous Waste Management, Atlanta, September 21–23, 1992.
13. J. M. Douglas and D. C. Woodcock, "Cost Diagrams and the Quick Screening of Process Alternatives," *Ind. Eng. Chem. Proc. Des. Dev.,* 24:970–976, 1985.
14. A. P. Rossiter, H. Klee, Jr., and R. D. Sung, "Process Integration Techniques for Pollution Prevention," American Chemical Society I&EC Special Symposium on Emerging Technologies for Hazardous Waste Management, Atlanta, September 21–23, 1992.
15. A. P. Rossiter and M. A. Rutkowski, "Process Integration for Wastewater Minimization," Southern States Annual Environmental Conference, Biloxi, MS, October 28, 1993.
16. R. F. Dunn and M. M. El-Halwagi, "Optimal Recycle/Reuse Policies for Minimizing the Wastes of Pulp and Paper Plants," *J. Environ. Sci. Health,* A28(1):217–234, 1993.
17. M. M. El-Halwagi, "Synthesis of Reverse-Osmosis Networks for Waste Minimization," *AIChE Journal,* 38(8):1185–1198, 1992.
18. R. Nath and J. D. Kumana, "NO_x Dispatching in Plant Utility Systems—Using Existing Tools," 14th Industrial Energy Technology Conference, Houston, April 22–23, 1992.

Pinch Analysis

5

Pinch Analysis in Pollution Prevention*

Prof. Bodo Linnhoff
Chairman and CEO
Linnhoff March International, Ltd.
Manchester, England

Introduction

In current industrial practice, most energy is obtained from the combustion of fuels. This results in unwanted by-products in the form of airborne emissions, notably NO_x, SO_x, CO_x, and particulates, which contribute to local pollution as soot and grime in industrial areas and to global pollution through acid rain and greenhouse gases. Energy conversion also creates solid and liquid wastes, such as ash residues and wastewater, following treatment for boilers and cooling towers. It is axiomatic that methods for reducing industrial energy demands must form a crucial element in any strategy for pollution prevention and waste minimization.

In a wide range of industries, pinch analysis is accepted as the method of choice for identifying medium- and long-term energy conservation opportunities to meet this need. It first attracted industrial interest in the 1970s, against the background of rising oil prices. An engineering approach, but one incorporating scientific rigor, pinch analysis quantified the potential for reducing energy consumption

---

*Figures in this chapter copyrighted by Linnhoff March. Used with permission.

through improved heat integration by design of improved heat-exchanger networks. Since then, pinch analysis has broadened its coverage to include reaction and separation systems in individual processes and entire production complexes. Today's techniques assess quickly and easily the capital and operating-cost implications of design options that reduce intrinsic energy consumption, determine the potential for capacity debottlenecking, improve operating flexibility, and, most recently, quantify the scope for reduced combustion-related emissions. In addition, the same basic concepts have been extended to mass-transfer problems to give similar insights into waste minimization, with significant environmental implications.

In this chapter the fundamentals and some extensions of pinch analysis are briefly reviewed. Chapters 6 to 8 detail how the concepts and principles are applied to industrial pollution prevention: through energy conservation in individual process units (Chap. 6), through global energy and emissions control for entire industrial complexes (Chap. 7), and by applying mass-transfer "water pinch" analysis to wastewater minimization (Chap. 8).

The Basics

Pinch analysis is based on rigorous thermodynamic principles. These are used to construct plots and perform simple calculations that yield powerful insights into heat flows through processes. The key concepts are *stream-based T-H plots* and are best explained in terms of two types of graphical representation: composite curves and grand composite curves.[1]

Stream *T-H* Plots

The starting point is a stream-by-stream breakdown of the heat sources and sinks within a process. This shows all heat sinks (cold streams) and all heat sources (hot streams) in a given process in terms of the enthalpy H and temperature T. In the first example there are only two streams, a and b (see Fig. 5.1a).

The interval boundary temperatures correspond to the start or end of a stream. In this case there are four, labeled T_1 through T_4. Next, the stream population between any two adjacent interval boundary temperatures is determined. Here stream a exists between T_1 and T_2; both streams a and b exist between T_2 and T_3; and stream b exists between T_3 and T_4. This quantifies the heating demand in each interval. If the specific heat capacities of all streams are constant, it is simple to calculate the heat load contribution each stream makes in each interval. The results are shown in Fig. 5.1b, in ascending temperature order from left to right. The single continuous curve represents all heat sinks

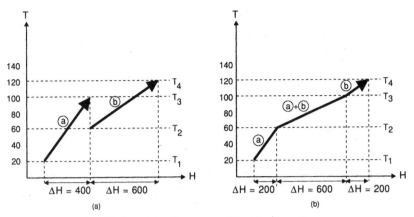

Figure 5.1 Stream T-H plots: (a) cold streams; (b) cold composite curve.

of the process as a function of heat load versus temperature. This plot is the cold composite curve.

By an analogous procedure, the process heat sources (hot streams) are combined to generate a hot composite curve. This represents all heat sources of the process as a function of heat load versus temperature.

The analysis is simplified by virtue of the facts that constant specific heats are assumed and that the number of streams is small. In practice, variable specific heats are accommodated by segmented linearization of data, and a large number of streams pose no practical problems with the computational power usually available.

Composite Curves, Energy Targets, and the Pinch

The two composite curves are conveniently plotted on the same axes in Fig. 5.2. Their positions are set such that at the closest vertical approach between them, the hot composite curve is ΔT_{min} higher than the cold composite curve, where ΔT_{min} is the minimum permissible temperature difference for heat transfer.

The combined plot is referred to as the *composite curves* and is a powerful tool for establishing energy targets. In the overlapping region of the curves, for every point on the cold composite curve there is a corresponding point on the hot composite curve directly above and at a temperature at least ΔT_{min} higher. Therefore, heat sources exist within the process to satisfy all heating requirements represented by this portion of the cold composite curve, with temperature driving forces of at least ΔT_{min}. The projection on the horizontal axis (X in Fig. 5.2) defines the scope for process-to-process heat recovery.

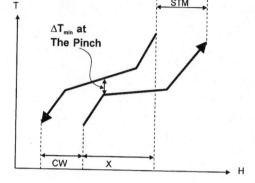

Figure 5.2 Combined composite curves.

The section of the hot composite curve to the left of the overlap region represents the quantity of heat that cannot be recovered in cold process streams. This has to be rejected to a cold utility, e.g., cooling water, labeled CW in Fig. 5.2. Similarly, the section of the cold composite curve to the right of the overlap region defines the quantity of utility heating (such as steam) that is required, labeled STM in Fig. 5.2. The quantity CW is the cold utility target for the process at the specified ΔT_{min}. The quantity STM is the corresponding hot utility target.

These targets represent realistically attainable goals, based on thermodynamic principles. They give the minimum amounts of utility cooling and heating that are necessary to drive the process, subject to maintaining a temperature driving force of at least ΔT_{min} in any network of heat exchangers, heaters, and coolers—yet to be designed—that services all sources and sinks in the given process.

The point in Fig. 5.2 at which the curves come closest together is the *pinch* point.[2] The vertical distance here between the curves is equal to ΔT_{min}.

The Pinch and Heat-Exchanger Network Design

Consider the characteristics of a *heat-exchanger network* (HEN) that is to achieve the energy targets, subject to a given ΔT_{min} requirement. By inspection of Fig. 5.2, we see that hot utilities should be used to heat cold streams above the pinch, and cold utilities should be used to cool hot streams below the pinch. Given that both hot utilities and hot streams are heat sources, and that hot streams below the pinch are not hot enough to heat cold streams above the pinch, if *any* hot stream above the pinch is used to heat a cold stream below the pinch, then the actual energy requirement will exceed the target (see Fig. 5.3). Any quantity Y of heat that is transferred across the pinch has to be replaced by an equal

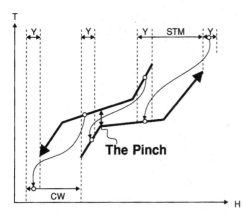

The Pinch

Figure 5.3 Cross-pinch heat transfer.

quantity of hot utility to satisfy the remaining heating requirements of cold streams above the pinch. Next, there is an equivalent increase in cold utility requirements. This leads to a fundamental axiom:

- Do not transfer any heat across the pinch.

From this are derived what have become known as the *three golden rules* of HEN design:

- Do not recover process heat across the pinch.
- Do not use hot utilities below the pinch.
- Do not use cold utilities above the pinch.

These rules are fundamental to the design of HENs for minimum energy requirements and are summarized in the pinch heat-transfer equation:

$$\text{Actual} = \text{target} + \text{XP}_{\text{proc}} + \text{XP}_{\text{hot below}} + \text{XP}_{\text{cold above}}$$

where actual = actual heat consumption of process
  target = minimum consumption
  XP_{proc} = heat transfer from hot process streams above the pinch to cold process streams below
  $\text{XP}_{\text{hot below}}$ = heat transfer from hot utilities to cold streams below the pinch
  $\text{XP}_{\text{cold above}}$ = heat transfer to cold utilities from hot streams above the pinch

The hot and cold composite curves represent the heat available in process hot streams and the heat required for process cold streams, respectively. Since they are separated by ΔT_{min} at the pinch, it follows that heat exchangers in the pinch region must approach ΔT_{min} at the pinch. This requirement and the golden rules above provide a systematic design procedure for the placement of exchangers in overall networks that guarantees the energy targets, the pinch design method.[3] For the purposes of design, the processes are divided into separate subsystems above and below the pinch.

The Grid Diagram

It is convenient for design purposes to have a representation for stream data and the pinch that also shows exchanger and utility placement in a simple graphical form. This is provided by the grid diagram in Fig. 5.4,[4] in which hot streams run left to right at the top, cold streams run right to left at the bottom, and the pinch position is indicated by a broken line at the division between the above- and below-the-pinch subsystems. Process exchangers are represented by circles on the streams matched joined by vertical lines, heaters by circles on cold streams and coolers by circles on hot streams. The grid makes it obvious when cross-pinch transfer occurs, and the design task becomes straightforward. Note that the grid allows the design and evolution of *any* network structure without the need to ever reposition a stream.

The Grand Composite Curve, Appropriate Placement, and Balanced Composite Curves

The hot and cold composite curves are representations of *process* heat sources and sinks. This information is now amalgamated into a single plot. After an allowance is made for the ΔT_{min} requirement in each temperature interval, the heat duties of all hot streams are offset against

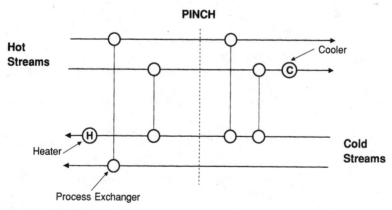

Figure 5.4 Grid diagram.

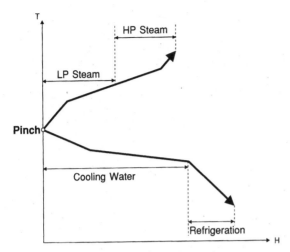

Figure 5.5   Grand composite curve with utility targets.

those of all cold streams.[1] The numerical procedure is called the *problem table*,[4] and the plot that results from it, Fig. 5.5, is the *grand composite curve.*[1]

The grand composite curve represents the net heating or cooling requirement of the process as a function of temperature. In other words, it depicts heating and cooling requirements at the process-utility interface. It enables the engineer to screen feasible options for the hot utility, e.g., furnace firing, hot oil, steam at various pressure levels, etc., and the cold utility, e.g., air cooling, water cooling, refrigeration at different levels, to satisfy the process requirements and to integrate heat engines and heat pumps with the process. This is the concept of *appropriate placement.*[5]

The utility and the process streams are next integrated into one plot, the balanced composite curve (Fig. 5.6), that shows the total heat balance of the system; and the exchangers and utilities are designed as a merged system, using the balanced grid (Fig. 5.6).[6]

Implications for Process Emissions

Pinch technology as described saw industrial use during the late 1970s and early 1980s. Although the purpose was primarily energy cost reduction studies, other benefits for capital cost reduction and the environment became apparent. Combustion-related atmospheric emissions, for example, SO_x, NO_x, and CO_x, are proportional to fuel firing rates, and consequently targets for these emissions can be established via hot utility targets.

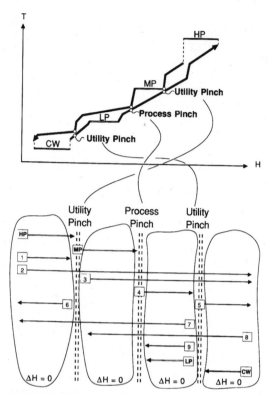

Figure 5.6 Balanced composite curves and balanced grid.

An early illustration of this analysis was a major campaign conducted by the German company BASF at its Ludwigshafen factory in the early 1980s.[7] Pinch methods were used to improve the energy efficiency of most of the processes on site. The objective was to save energy and reduce related operating costs. However, a review of the results clearly demonstrated that major environmental benefits were also obtained, with significant reductions in combustion-related atmospheric emissions. Wastewater emissions were also decreased, as less water treatment was required for steam and cooling water. The results of this campaign are discussed further in Chap. 6.

Findings such as these paved the way for subsequent similar studies and campaigns[8,9] (see also Chap. 23) using pinch analysis specifically for the combined objectives of saving energy and reducing emissions. In recent developments, outlined briefly below and covered in detail in later chapters, pinch methods are used for reducing emissions from total production sites as opposed to individual processes.

Extensions of Pinch Analysis

Since the mid-1980s, the scope of pinch analysis has broadened considerably; but it is still rooted firmly in thermodynamics, and all developments maintain the key strategy of setting targets prior to design. It is now an overall methodology for what is sometimes referred to as *process synthesis,* including the design of reactors, separators, furnaces, combined heat and power systems, etc., in the process context. It is the method of choice for identifying a wide range of process improvement options for the heart of chemical processes, or the heat and material balance. Objectives can include energy cost reduction, capital cost reduction, capacity increase, yield improvement, and, of course, emissions reduction.[10] It is used equally for retrofits and for grass-roots design, and now for planning in both continuous and batch industries. An up-to-date overview was given recently.[11]

Specific developments have made pinch analysis a versatile and effective weapon in the attack on combustion-related pollution:

Supertargeting

It is possible to predict from the composite curves the overall target heat-transfer surface area and the target capital cost. Based on the target energy cost and target capital cost, the target overall cost is predicted for any given ΔT_{min}. This concept of *supertargeting*[12] (see Fig. 5.7) enables the designer to optimize ΔT_{min} *prior to* design. In other words, pinch analysis addresses not only the energy cost but also the capital cost. It guides the overall optimization of the total cost of integrated systems at the targeting stage.

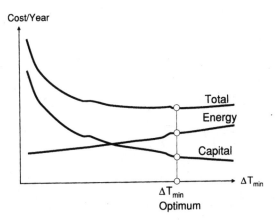

Figure 5.7   Supertargeting.

The plus/minus principle

Process changes are changes to the basic process heat and material balance which may or may not include the chemistry of the process. The *onion diagram*[1] is used to represent the hierarchy of chemical process design. The start is from reactors converting feeds to products. Once the reactor design is settled, the separation needs are known and the separators can be designed. Once the separator design is settled, the mass flows, vapor/liquid equilibrium, temperatures, pressures, and flow rates are known, and the heat-exchanger network can be designed. Prior to designing the heat-exchanger network, the basic heat and material balance is defined and the composite curves representing the base-case process are constructed.

The *plus/minus principle*[13] (see Fig. 5.8) uses the composite curves to identify process changes by going backward into the onion, based on the recognition that the hot utility target is reduced if the hot stream enthalpy balance above the pinch is increased, or the cold stream enthalpy balance above the pinch is reduced. It is essentially a generalization of appropriate placement, for desirable process changes (relating to distillation, reaction, etc.). It gives the designer guidelines as to what changes are beneficial for energy and/or capital costs.

Column profiles

Conventional pinch analysis concerns heat exchange between streams within a process. Column composite curves and column grand composite curves enable a similar analysis "inside" distillation columns, by using a plot of reboil and condensing temperatures called the *column profile*[14] (see Fig. 5.9). Similar to a grand composite curve, the column profile indicates at what temperatures heat needs to be supplied and rejected up and down the column. Not all heat needs to be supplied at the reboil temperature; some heat can be supplied at lower temperature. Not all heat needs to be removed at condensing temperature; partial heat removal at higher temperatures may be appropriate.

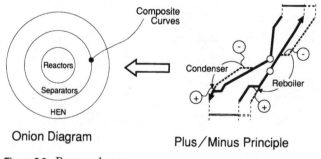

Figure 5.8 Process changes.

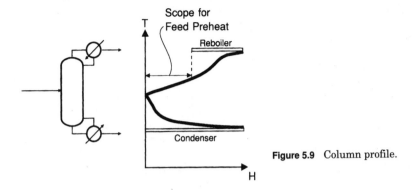

Figure 5.9 Column profile.

Shaftwork targeting

The grand composite curve establishes targets for utilities for any given process. If the plot is generated by using the Carnot factor as the vertical axis instead of temperature, specific areas on the plot represent changes in exergy or available energy, which is equivalent to ideal shaft work. This exergy grand composite curve[15] (see Fig. 5.10) identifies process improvements in the design of processes that include refrigeration systems. The effects of process changes can readily be evaluated. This greatly simplifies the design of below-ambient processes.[16,17]

Overall Pollution Prevention

The concepts described above have been used extensively in the last 5 to 10 years in industrial pollution prevention through energy conservation in individual process units. Two other developments have had a

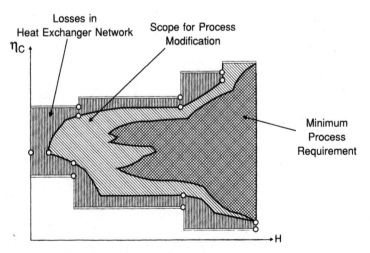

Figure 5.10 Exergy grand composite curve.

major impact on overall pollution prevention on entire sites. Outlined briefly here, they are covered in depth in Chaps. 7 and 8.

Total Site Integration

Combustion-related pollution is typically considered a local problem, related to the boilers on a manufacturing site. In reality, boiler emissions are due to energy demands and losses throughout the facility. Indeed, the problem is even more global. Most large process installations generate, import, or export power with steam turbines, gas turbines, or other prime movers; and load variations will invariably have an impact on the power required from electricity suppliers, thereby affecting emissions from the public utilities.

When a study of such a large, multiprocess production facility is required, the traditional pinch analysis techniques for single processes become unwieldy. A new concept addresses this problem, setting site-wide targets for combustion-related emissions[18] (see Fig. 5.11).

In terms of the onion diagram (Fig. 5.8), reactors and separators form the inner core, surrounded by the heat recovery layer and then the utilities layer. The composite curves link the core and the recovery layer; the grand composite curves link the core directly to the utility layer. Together these concepts tell the engineer what options there are in process design and the consequent utility choices and allow iteration

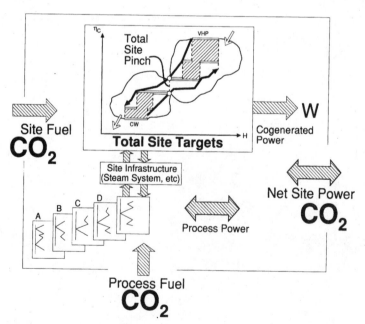

Figure 5.11 Emissions targeting.

back to the core for optimization. This simple translation of process designs into utility mix provides the key to the analysis of a total site.

In a typical site with several production processes, some processes have their own utilities, but others do not. There is usually a common set of central utilities, on site and external, using imported fuel and/or direct power.

By using pinch analysis, each process can be represented by its optimized grand composite curve. These curves are combined into a set of *site source-sink profiles*. The source profile depicts the heat remaining after all practical process heat integration. It is the net heat available from all processes, in terms of the quantity of energy and the temperature level, and is analogous to the hot composite curve for a single process. Similarly, the sink profile represents the heat required after such integration, the net heat demand, and is analogous to the cold composite curve. These total site profiles give a simultaneous view of surplus heat from, and heat deficit for, all site processes.

Utilities are placed in the total site profiles, with the various utility levels specified, to give explicit cogeneration and fuel targets for the overall site. The heat loads for different site steam levels, a turbine system for cogeneration, and central boiler capacity and cooling demands are all obtainable. From the fuel profile and from the fuel composition, emission levels are derived. Simultaneous targets are therefore set for site fuel, power export/import, cooling, and emissions.

Once the needs for on-site fuel and imported energy are balanced, it is possible to target global emissions, using an overall site targeting procedure. Economic decisions and trade-offs are made for process changes, fuel and steam demands, and infrastructure changes and are used to set targets. The decisions are tested in the global context, revised as required, and the procedure is repeated to establish a set of planning scenarios meeting different criteria. As the targets meet specific conditions and constraints, the designer can with confidence produce strategies for site development packages that take into account long- and short-term investment/benefit functions. Arguments for local and global control measures can be rationally assessed.

This approach has been successfully demonstrated in studies of both new and existing multiunit process plants. It is covered in detail in Chap. 7.

Mass-Transfer Water Pinch Analysis

Another type of industrial pollution is effluent, in particular liquid waste from the water treatment plant. Ever stricter discharge regulations drive up treatment costs; that means capital expenditure with little or no productive return. Effluent reduction has direct economic advantages; the consequent reduction of freshwater intake also

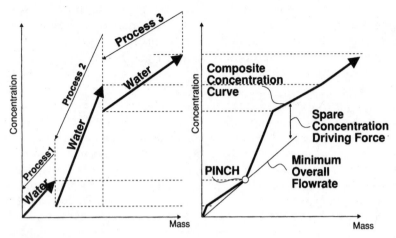

Figure 5.12   Water pinch.

reduces operating costs. By taking advantage of certain parallels between the principles of heat and mass transfer, the systematic design procedures of pinch analysis have been extended to address these problems[19] (see Fig. 5.12).

Process streams are brought into contact with water streams to reduce the contaminant level in the process. The goal is to reduce water throughput with no detrimental effects. The options are the reuse of water in different operations, regeneration and reuse, or regeneration and recycling. Freshwater and wastewater flows are reduced in each case, and in the latter two cases the contaminant load of the wastewater is also reduced.

The idea is to plot the process data as contaminant concentration-mass curves (compare temperature-enthalpy curves). To maximize the water reuse (i.e., recovery) potential, the highest acceptable inlet concentration is needed. Specifying the maximum acceptable outlet concentration minimizes the water flow rate. The plots so produced are *limiting water profiles*; any water supply line below these profiles can satisfy the process cleanup requirement. Constraints are built in.

Multiple-stream data are combined over concentration intervals to produce the concentration composite curves. An overall freshwater supply line is plotted that touches this curve at the most limiting intermediate point, which is the *water pinch*. The slope of this line maximizes outlet concentration and targets minimum freshwater use and wastewater flow rate.

In practical systems there may be multiple contaminants and multiple constraints. By using concentration shift techniques, equivalent analytical procedures are applied. Target designs are achieved by using systematic methods. The designer has insight into alternative structures, essential and optional features. Process changes that improve the target and critical areas of design are made obvious.

Better integration through mass-transfer water pinch analysis has considerably reduced discharges from facilities where conventional water reduction programs had already made significant savings.

The methodology of water pinch analysis and its applications are covered in detail in Chap. 8.

References

1. B. Linnhoff, D. W. Townsend, D. Boland, G. F. Hewitt, B. E. A. Thomas, A. R. Guy, and R. H. Marsland, *User Guide on Process Integration for the Efficient Use of Energy*, IChemE, Rugby, UK, 1982.
2. B. Linnhoff, D. R. Mason, and I. Wardle, "Understanding Heat Exchanger Networks," *Comp. Chem. Eng.*, 3:295–302, 1979.
3. B. Linnhoff and E. Hindmarsh, "The Pinch Design Method of Heat Exchanger Networks," *Chem. Eng. Sci.*, 38(5):745–763, 1983.
4. B. Linnhoff and J. R. Flower, "Synthesis of Heat Exchanger Networks: Part I: Systematic Generation of Energy Optimal Networks," *AIChE Journal*, 24(4):633–642, 1978, and "Part II: Evolutionary Generation of Networks with Various Criteria of Optimality," *AIChE Journal*, 24(4):642–654, 1978.
5. D. W. Townsend and B. Linnhoff, "Heat and Power Networks in Process Design, Part I: Criteria for Placement of Heat Engines and Heat Pumps in Process Networks," *AIChE Journal*, 29(5):742–748, 1983; "Part II: Design Procedure for Equipment Selection and Process Matching," *AIChE Journal*, 29(5):748–771, 1983.
6. B. Linnhoff and J. de Leur, "Appropriate Placement of Furnaces in the Integrated Process," paper presented at IChemE Symposium *Understanding Process Integration II*, March 22–23, 1988 University of Manchester Institute of Science and Technology, Manchester, UK.
7. H. Körner, "Optimal Use of Energy in the Chemical Industry," *Chem. Ing. Tech.*, 60(7):511–518, 1988.
8. S. Morgan, "Use Process Integration to Improve Process Designs and the Design Process," *Chem. Eng. Progress*, September 1992, pp. 62–68.
9. Y. Natori, "Managing the Implementation of Pinch Technology in a Large Company," paper presented at the International Energy Agency Workshop on Process Integration, Gothenburg, Sweden, January 28–29, 1992.
10. R. Smith, E. A. Petela, and H. D. Spriggs, "Minimization of Environmental Emissions through Improved Process Integration," *Heat Recov. Syst. CHP*, 10(4):329–339, 1990.
11. B. Linnhoff, "Pinch Analysis: Building on a Decade of Progress," *Chem. Eng. Progress*, August 1994.
12. B. Linnhoff and S. Ahmad, "Cost Optimum Heat Exchanger Networks, Part 1: Minimum Energy and Capital Using Simple Models for Capital Cost," *Comp. Chem. Eng.*, 14(7):729–750, 1990.
13. B. Linnhoff and D. R. Vredeveld, "Pinch Technology Has Come of Age," *Chem. Eng. Progress*, July 1984, pp. 33–40.
14. V. R. Dhole and B. Linnhoff, "Distillation Column Targets," *Comp. Chem. Eng.*, 17(5/6):549–560, 1993.
15. B. Linnhoff, "Pinch Technology for the Synthesis of Optimal Heat and Power Systems," *Trans. ASME, J. Energy Res. Tech.*, 111(3):137–147, September 1989.
16. B. Linnhoff and V. R. Dhole, "Shaftwork Targets for Low Temperature Process Design," *Chem. Eng. Sci.*, 47(8):2081–2091, 1992.
17. V. R. Dhole and B. Linnhoff, "Overall Design of Subambient Plants," paper presented at ESCAPE-III Conference, Graz, Austria, July 1993. Also printed in *Comp. Chem. Eng.*, 18:s105–s111, 1994.
18. V. R. Dhole and B. Linnhoff, "Total Site Targets for Fuel, Co-generation, Emissions, and Cooling," *Comp. Chem. Eng.*, 17:s101–s109, 1993.
19. Y. P. Wang and R. Smith, "Wastewater Minimization," *Chem. Eng. Sci.*, 49(7):981–1006, 1994.

6

Pinch Analysis to Reduce Individual Process Emissions

Eric A. Petela
Linnhoff March, Ltd.
Knutsford, England

Introduction

As discussed in the previous chapter, pinch analysis first attracted attention as a means for reducing industrial energy consumption in the late 1970s and early 1980s. The techniques rapidly became established as an effective means of helping with a new problem facing the process industries at that time—rapidly rising energy costs and their impact on profitability. Today, in an increasingly environmentally conscious world, pinch analysis is being used to solve a problem of the 1990s— how to make the process plant environmentally more acceptable, a problem of waste minimization and pollution prevention. In this chapter the discussion focuses on the use of pinch methods to minimize emissions from individual processes. The adaptation of these techniques to deal with total sites (including multiple processes and infrastructure issues) is addressed in the next chapter.

Pinch analysis provides a systematic mechanism for identifying opportunities to reduce combustion-related emissions (especially airborne NO_x, SO_x, CO_2, and particulates). As discussed in the previous chapter, these opportunities arise not only from better heat-exchanger network designs, but also from improved overall process designs.

Exploiting these opportunities makes a process as efficient as is technically and economically possible in its use of energy—an example of waste minimization. At the same time, operating costs are lower, by virtue of the reduced energy requirements for the process. Moreover, the need for handling combustion wastes (e.g., ash collection, NO_x and SO_x removal, etc.) is also reduced, saving additional capital and operating costs (see also Chap. 18).

Reducing Emissions by Improving Heat Integration

The fact that pinch analysis is an effective tool for reducing both energy use and emissions is emphatically demonstrated by Körner.[1] He reports the results of a major energy-saving campaign at BASF's Ludwigshafen factory in the early 1980s. Pinch analysis, then in its infancy, formed a major part of the campaign, and pinch methods were used to improve the heat integration of most of the major processes on the site. A simple payback of 1 year or less, based on the energy-saving potential alone, was required for each project. The total energy savings (as fuel) amounted to 790 MW, and this was achieved together with increased production.

The improvements in energy efficiency in the individual processes inevitably reduced the fuel firing requirements at the factory. Combustion-related airborne emissions and ash residues therefore decreased. Alongside these benefits, wastewater discharges were reduced, since less water treatment was required for steam and cooling water. The emission reductions are summarized here:

Carbon dioxide	218 t/h
Sulfur dioxide	1.4 t/h
Nitrogen oxides	0.7 t/h
Ash	21 kg/h
Carbon monoxide	7 kg/h
Wastewater from water treatment	70 t/h

Significantly, these benefits were obtained "automatically" when the energy efficiency of production processes improved. Furthermore, they were accompanied by a reduction in energy-related operating costs. The BASF experience has been followed by many similar examples of campaigns where pinch analysis has been used with the combined objectives of energy efficiency and waste minimization (see, for example, Chap. 23).

Reducing Emissions by Improving Process Design

Pinch analysis is not just a means of identifying energy-saving opportunities; it also provides a detailed understanding of the mass and energy flows in a process. Figure 6.1 illustrates the fact that the mass and energy balances of a process are intimately linked. Raw materials are converted into products and waste but it is the energy systems which drive the process. Both balances need to be considered in any review of waste minimization.

Waste minimization needs to start at the heart of the process—in the reactor.[2] The reactor is often the major source of waste. Correct selection of reactor type and conditions can reduce the formation of unwanted by-products.[3] Once a reactor system has been established, the next major area for review is the separation system. If the separation system in a process can be made more efficient such that useful material can be separated and recycled, then this will reduce both raw material costs and effluent disposal costs.

Energy is an important aspect of both reaction and separation systems. Most reactions require either heat input or heat removal, and these needs can have a major impact on both the design of the reactor and the energy balance of the process. The separation systems used by industry are driven by energy in one form or another. Operations such as distillation and evaporation require heat input while others, such as condensation and crystallization, require the removal of heat.

Wherever heat input or heat removal occurs, pinch analysis plays a role in evaluating and optimizing the system. The trade-offs can be complex. Capital, raw materials, waste disposal, and energy often have

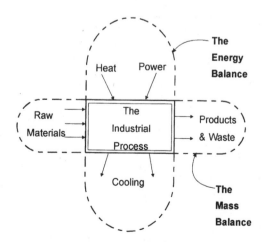

Figure 6.1   Two fundamental balances are important to process design.

to be considered simultaneously. Without a systematic tool such as pinch analysis, such trade-offs are often impossible to perform. This is illustrated by the following examples, which are based on recent projects with which the author has been associated.

Example 1: Combustion-related emissions

Figure 6.2 shows the grand composite curve of a petrochemical process. As discussed in Chap. 5, the grand composite curve represents the net heating and cooling requirements of the process as a function of temperature. It can therefore be used to match utility duties (e.g., furnace firing, steam at various pressure levels, cooling water, and refrigeration) to the needs of the process.

In this example the process requires heating to a temperature which is higher than can be obtained by steam, so a furnace must be used. The flue gas profile is matched to the grand composite curve, keeping the flue gas temperature above the process temperature at all points. The minimum flue gas flow is determined from the reciprocal of the flue gas gradient.

The process has a relatively low pinch temperature. However, the flue gas profile in Fig. 6.2 is not limited by the process pinch. There is a "knee" in the grand composite curve which creates a process/utility pinch. This reduces the gradient of the flue gas profile and hence

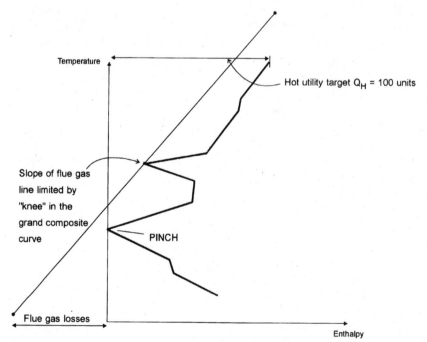

Figure 6.2   Grand composite curve of a petrochemical process.

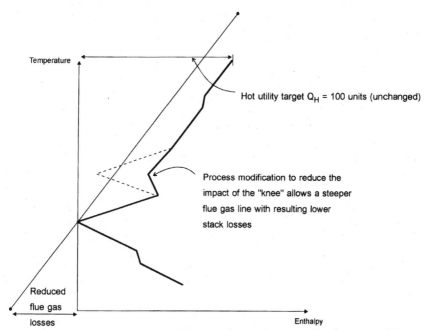

Figure 6.3   Grand composite curve of the petrochemical process after process modification.

increases the required flue gas flow. What this means in practice is that most of the heat recovered from the flue gas in the furnace would have to be transferred to high-temperature streams. The flue gas therefore leaves the stack hot, resulting in a poor furnace efficiency.

An understanding of this phenomenon allowed modifications to be made to the process to eliminate the knee by applying the *plus/minus principle* (see Chap. 5) around the utility pinch. This involved adjusting temperatures and heat loads in the separation system. The resulting grand composite curve is shown in Fig. 6.3. This allows a steeper flue gas profile to be used, leading to reduced fuel firing and emissions targets. The design of the furnace corresponding to this profile would include at least one convection bank at a relatively low temperature, enabling the flue gas to be cooled to the process pinch. Consequently the stack temperature is lowered, the efficiency of the furnace increases, and emissions are reduced.

Example 2: Liquid and solid emissions

While there is general acceptance that pinch analysis can be used to reduce gaseous emissions associated with the provision of energy, what is less widely recognized is the ability of pinch analysis to identify improvements in other areas.

The overall insight provided by pinch analysis can often be an extremely effective means of identifying such improvements. Consider the following example:

Figure 6.4 shows the final stage of a production process. A concentrated solution of an inorganic salt is fed to a rotary, gas-fired calcination unit. In the original design, the gases leaving the calcination unit are sent to the atmosphere. Since this exhaust contains "fines," it was deemed necessary to add a cleanup facility. After a thorough evaluation of cleanup options, a water scrubber was chosen. The proposed scheme is shown in Fig. 6.5.

The effluent from the scrubber is a weak solution of product in water, and it was decided that this should be recovered. The problem, however, was that this solution would be too dilute to recycle to the calcination unit. Therefore, concentration of the solution would be required.

Conversation with equipment suppliers produced a design for a three-stage evaporator to give high energy efficiency. The proposed

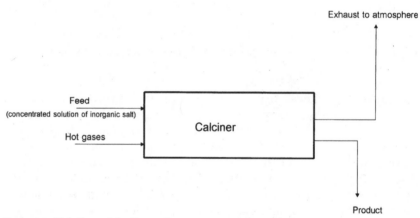

Figure 6.4 Existing calcination unit.

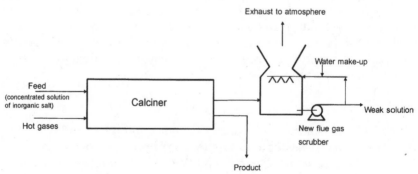

Figure 6.5 Proposed cleanup facility on calcination unit.

evaporator, while energy-efficient, would require a high level of capital investment, making the whole cleanup operation very expensive. At this point, questions started to be raised about the validity of the original proposal of water scrubbing as a means of cleaning the exhaust. Perhaps a bag filter would be more appropriate. It was soon realized, however, that the fines recovered by filtration would be too small to simply blend into the final product and would have to be redissolved, thereby giving rise to the same problem: How to recycle? At this point, it was decided to readdress the whole problem, and pinch analysis was identified as a means of obtaining a proper overview of the issues involved.

The grand composite curve of the total process is shown in Fig. 6.6. This shows a process pinch at 50°C and a large demand for heating at 75°C. The heating requirement was being supplied as low-pressure steam from the boiler plant. The analysis highlighted the potential to install an "appropriately placed" evaporation plant, as shown in Fig. 6.7. The final solution is shown in Fig. 6.8. Steam from the boiler plant is fed in part to the process, but in part to a new single-stage (and hence low-capital-cost) evaporation plant. The evaporator concentrates the effluent prior to recycling to the calcination unit. The operating pressure of the evaporator is not atmospheric. Rather, it is chosen at such

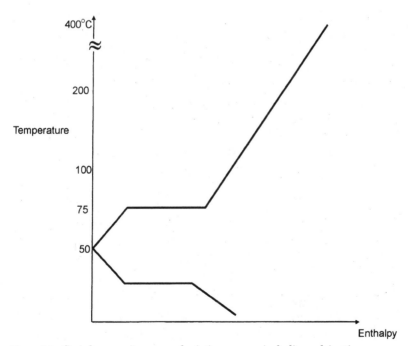

Figure 6.6 Grand composite curve of existing process including calcination.

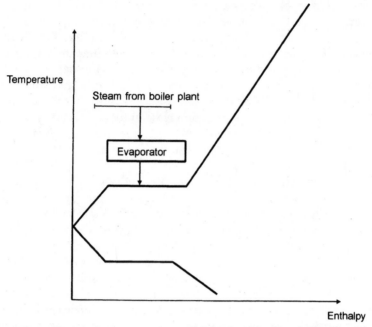

Figure 6.7   Proposed evaporator appropriately placed in the overall process.

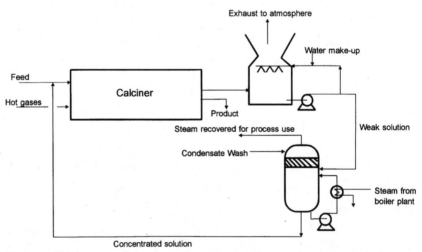

Figure 6.8   Final process flow scheme incorporating flue gas cleaning and single-stage evaporation.

a level that the steam produced from evaporation is at a temperature which is high enough that it can be recovered and used to supply process heating duties. These duties were previously supplied by steam from the boiler plant.

The overall steam consumption in this modified design (Fig. 6.8) is only very slightly higher than that of the original process design (Fig. 6.4). As a result, the evaporation plant runs at near-zero operating cost. Moreover, the capital cost for this simple unit was significantly less than that for the three-stage system originally proposed.

This example demonstrates how pinch analysis can be used most effectively to reduce waste, to aid in the selection of the most appropriate form of effluent treatment options, and to make emissions reduction more cost-effective.

Pinch Analysis and Waste Minimization in Perspective

A structured approach to reducing emissions by waste minimization has been proposed by Smith and Petela.[2] They argue that the heart of any process is the reactor, that this is often the place where wastes are first generated, and that hence the reactor is the place where effort should first be directed as part of any waste minimization program. Thereafter, attention should be focused on the separation systems, heat recovery systems, and finally the utilities plant. Similar hierarchies have been proposed by Douglas[4] and by Rossiter et al.[5] (see Chap. 12). Many different factors have to be considered at each stage—process chemistry, types of equipment to be selected, operating conditions, and safety considerations, to mention but a few.

Pinch analysis does not answer all the questions. However, at each stage it has a part to play. It assists in identifying opportunities, evaluating options, and maximizing the return on investment. It clearly has an important role to play in the design of heat recovery systems and utilities; but it also has applications in optimizing reaction and separation systems, as illustrated in this chapter. Finally, it has also been demonstrated that when effluent cleanup facilities are to be added to a process, once again pinch analysis can ensure the most cost-effective solution.

References

1. H. Körner, "Optimal Use of Energy in the Chemical Industry," *Chem. Ing. Tech.,* 60(7):511–518, 1988.
2. R. Smith and E. Petela, "Waste Minimization in the Process Industries, Part 1: The Problem," *The Chem. Engineer,* October 31, 1991, pp. 24–25.
3. R. Smith and E. Petela, "Waste Minimization in the Process Industries, Part 2: Reactors," *The Chem. Engineer,* December 12, 1991, pp. 17–23.
4. J. M. Douglas, "Process Synthesis for Waste Minimization," *Ind. Eng. Chem. Res.,* 31:238–243, 1992.
5. A. P. Rossiter, H. D. Spriggs, and H. Klee, Jr., "Apply Process Integration to Waste Minimization," *CEP,* 89(1):30–36, January 1993.

Total Site Integration for Reducing Process Emissions*

Vikas R. Dhole, Ph.D.

Linnhoff March, Ltd.
Knutsford, England

Introduction

Combustion-related pollution (for example, NO_x, SO_x, CO_2, particulates) is typically considered a local problem, related to the boilers and furnaces on a manufacturing site. In reality, these emissions are due to energy demands and losses throughout the facility. Indeed, the problem is even more global. Because most large process plants generate at least some of their power with steam turbines, gas turbines, or other prime movers, load variations will have an impact on power import from electricity suppliers. This in turn affects emissions from public utilities. These interactions are illustrated in Fig. 7.1.

Pinch analysis assesses the minimum practical energy needs for a process and achieves these through a systematic methodology of targeting followed by design. In this chapter we summarize recent develop-

---

*Figures in this chapter copyrighted by Linnhoff March. Used with permission.

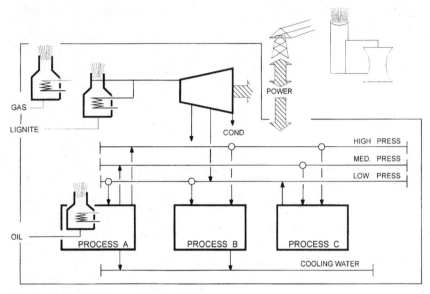

Figure 7.1 Schematic of a typical total site.

ments that extend the methodology to overall sites, i.e., multiple process-es linked by a common central utility system. Prior to design the engi-neer can target fuel, cogeneration, emissions (fuel-related), and cooling for options related to process or utility modifications or expansions. The targets help the designer set an overall plan for sitewide reduction in energy and emissions for a given investment scenario. Selected options can then be translated to designs. This is *total site integration.*

A Total Site

Figure 7.1 shows a schematic of a typical process industry site involv-ing several production processes. The production processes, some with their own utilities, are served by a common central utility system. The utility system consumes fuel (e.g., gas and lignite), generates power, and supplies the necessary process steam through several steam mains. There is both consumption and recovery of process steam via the steam mains. Process furnaces also consume fuel (e.g., oil). The site imports or exports power to balance the on-site power generation. The process heating and cooling demands and the cogeneration potential dictate the sitewide fuel demand via the utility system. The fuel-relat-ed emissions are directly related to the fuel consumption and the type of fuel used.

Usually, the individual production processes are designed and imple-mented on a one-off basis by company engineers or external contractors;

central services are controlled by a different department. A number of different business units may occupy different subsections of the site and operate independently. The site infrastructure therefore usually suffers from inadequate development overview. To rectify it requires a simultaneous approach to consider individual process issues alongside sitewide utility planning in the single context of making products from feedstocks with minimum use of capital and energy in all forms. Total site integration as discussed in this chapter provides an aid to the sitewide planning for reduction in energy and fuel-related emissions.

Total Site Profiles

The heating and cooling requirements of individual processes are represented by their respective grand composite curves; these lay open the process-utility interfaces for a single process. However, for a system involving several processes, the grand composite curve of each process will suggest different steam levels and loads. The identification of the correct compromises in steam levels on a sitewide basis is part of the new development.

The procedure for total site integration is based on generation of thermal profiles for the entire site, called *total site profiles*.[1] These are constructed from the grand composite curves of all the processes on the site. Figure 7.2 illustrates the construction with a two-process example.

Prior to the construction of the total site profiles, the grand composite curves of the individual processes are first modified in two ways:

1. First we need to isolate the parts of the grand composite curves that are satisfied by the central utility system. The nonmonotonic parts, or the *pockets,* are "sealed off" through vertical lines. Also the parts of the curve directly satisfied by a noncentral utility (such as furnace heating within a process) are excluded from the analysis. The remaining parts are the net heat source and sink elements of the curve to be satisfied by the central utility system.

2. The source and sink elements are shifted by $\Delta T_{min}/2$ as shown. Source element temperatures are reduced; sink element temperatures are increased. This is to ensure a minimum temperature difference ΔT_{min} between the steam mains and the process duties.

In the case where the processes have different ΔT_{min} values, the individual source and sink elements will be shifted by $\Delta T_{min}/2$ corresponding to the ΔT_{min} of the process.

The total site profiles are then constructed from the shifted source and sink elements as shown in Fig. 7.2. The construction simply involves the generation of *composite curves* from the shifted source and

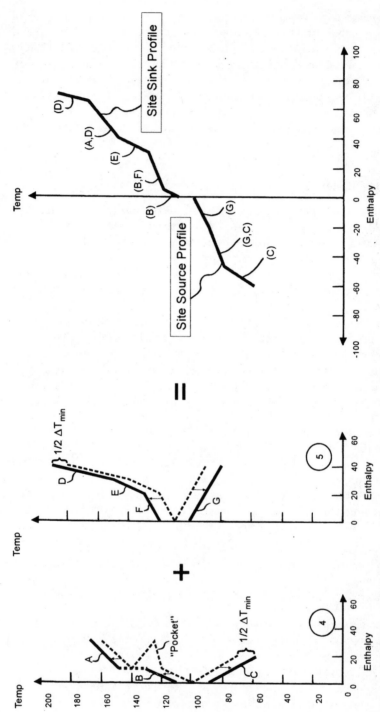

Figure 7.2 Constructing total site profiles from process grand composite curves.

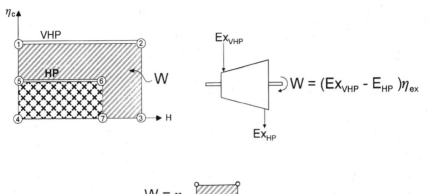

Figure 7.3 Shaftwork targeting for cogeneration.

sink elements. The composite of the source elements is called the *site source profile* while that of the sink elements is called the *site sink profile*. For illustrative purposes, the individual sink elements (*A, B, D, E,* and *F*) and source elements (*C* and *G*) are highlighted in Fig. 7.2.

Figure 7.3 illustrates an approach for setting cogeneration targets.[1] This involves a different choice of axes whereby the temperature scale on the vertical axis is replaced by the Carnot factor, while the horizontal axis remains the enthalpy (Fig. 7.3). Consider a simple very high-pressure (VHP) to high-pressure (HP) turbine. The exergy (ideal shaftwork equivalent) of the VHP steam is given by the shaded area 1-2-3-4, while the exergy of the HP steam is given by the shaded area 4-5-6-7. The net ideal shaftwork obtainable from the VHP to HP turbine equals the exergy difference, in other words, the shaded area 1-2-3-7-6-5. The real shaftwork is proportional to the ideal shaftwork, if we take into account the exergetic efficiency of the turbine system η_{ex}, thus setting the target for cogeneration.[1]

Figure 7.4 illustrates sitewide utilities targeting by using total site profiles. The figure shows a utilities schematic and the corresponding total site profiles with the utilities placed appropriately. As mentioned earlier, the vertical axis is changed from temperature to the Carnot factor to allow direct targeting for cogeneration. The target for medium-pressure (MP) steam generation from all the processes is *D*. This partly satisfies the MP demand. The remaining targeted MP demand *C* is provided by the turbine system. The targeted HP steam demand for all the processes is *B*. Thus the turbine system needs to satisfy MP steam demand *C* and HP steam demand *B*. Since the vertical axis is the Carnot factor, the shaded area is proportional to the cogeneration *W*.

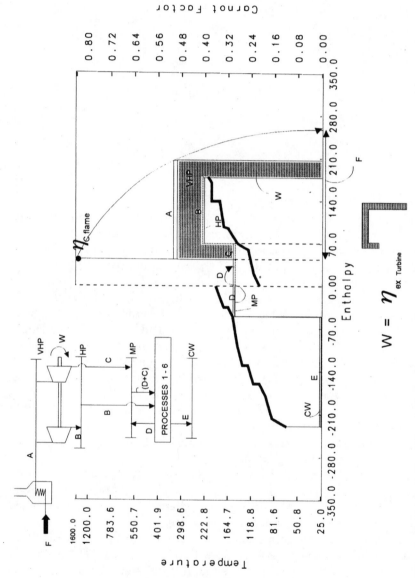

Figure 7.4 Total site targeting for fuel, cogeneration, emissions, and cooling.

$$W = \eta_{ex\ Turbine}$$

The generating steam requirement A at the VHP level can be identified. The VHP load sets the target for the fuel requirement of the boiler F as the horizontal projection of the fuel profile on the enthalpy axis. From a knowledge of the fuel composition and performance characteristics of combustion equipment, the quantity of each fuel-related emission (say, NO_x, SO_x, CO_2, particulates) can be determined as a function of the fuel consumption. By combining this information with the boiler fuel target, targets can now be set for emissions.

The cooling demand for the total site as identified by the total site profiles is E. Thus, by starting from individual process grand composite curves, simultaneous targets are set for site fuel, cogeneration, site emissions, and cooling demand.

Case Study: Site Expansion

The use of total site profiles is illustrated by the following example, which relates to a site expansion project.[1] The example shows how the procedure can identify options that result in not only reduced emissions, but also capital cost savings and lower operating costs than conventional solutions.

Figure 7.5 shows the example site with six processes as operating prior to the expansion project. The existing utility scheme and the total site profiles are shown. The high-pressure steam main receives steam from the central turbine system (190 MW). The medium-pressure main is not connected to the turbine system. At the MP level, the steam generation from processes and steam consumption by processes are in balance (30 MW each). The total site profiles construction reveals an overall cross-pinch heat transfer (in all processes and the utility system) of 40 MW. The site fuel consumption is 282 MW, and the cogeneration is 23 MW.

A site expansion is planned that involves the introduction of a new process into the existing site. Figure 7.6 shows the proposed expansion plan, with total site profiles and a utility sketch. Both the source and the sink profiles are enlarged as a result of the expansion.

The initial study subjected the new process to careful individual optimization, resulting in a minimum energy requirement of 70 MW of MP steam. This steam is to be supplied via a *new turbine* that links into the existing VHP and MP mains. To generate the additional VHP steam (from 213 to 290 MW), a *new boiler* is required. Also, an *expansion* of the cooling water system is necessary (from 240 to 280 MW).

However, the total site profiles reveal additional information. Since the profiles change significantly as a result of the expansion, MP steam is no longer appropriate: It does not exploit the driving forces available

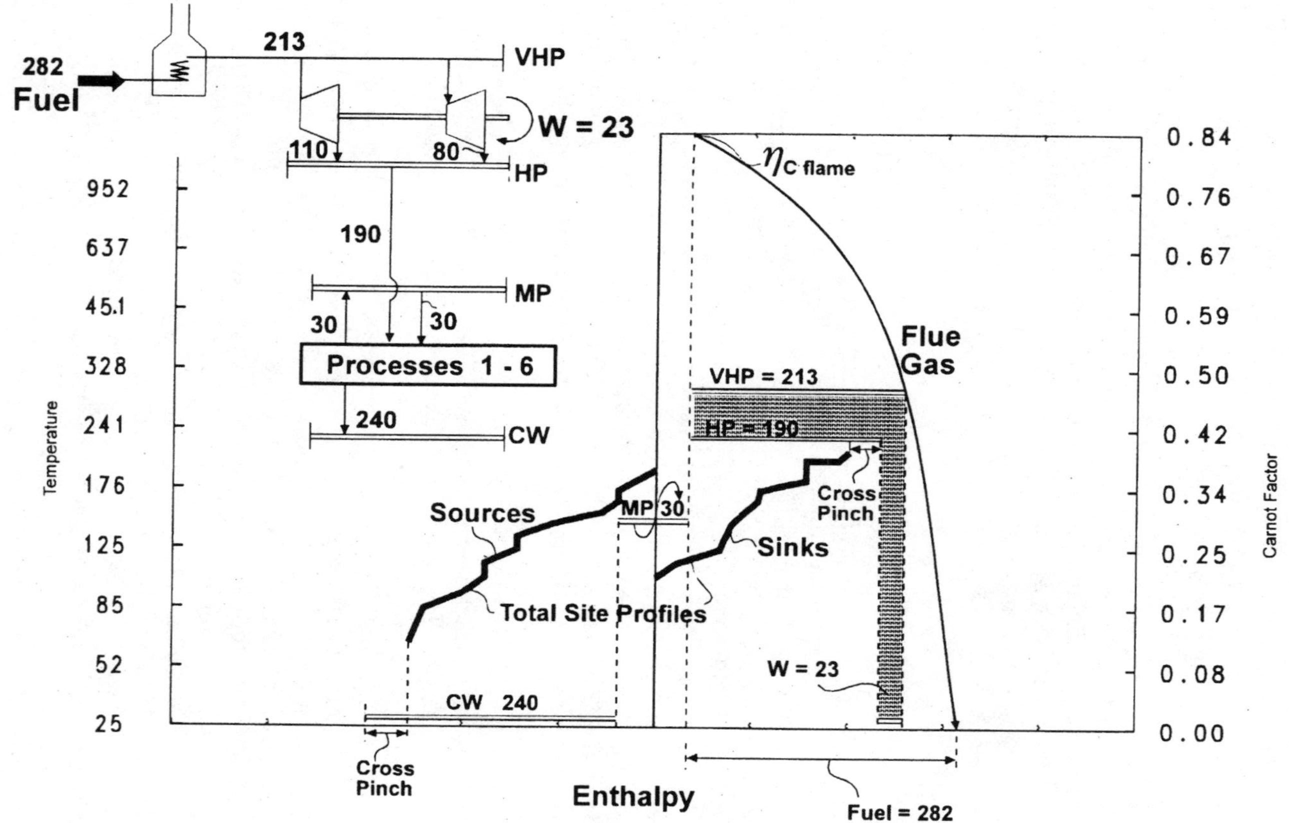

Figure 7.5 Existing site.

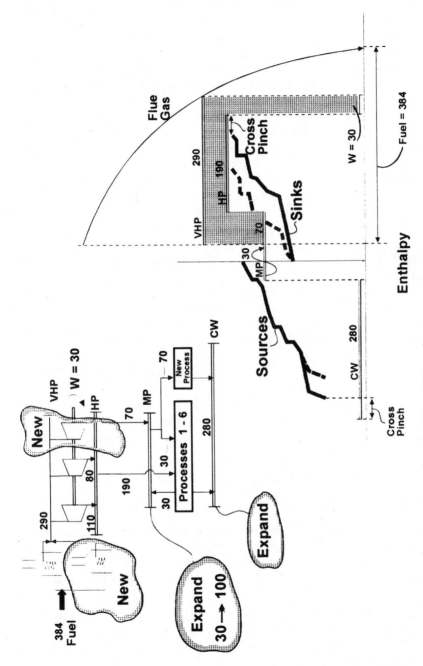

Figure 7.6 Proposed expansion of the site involving addition of a new process.

between the steam levels and the site sink profile. Pinch analysis indicates that MP steam pressure needs to be reduced. This is implemented in Fig. 7.7. The reduction in the MP steam pressure is selected such that MP steam generation is in balance with its use, *and therefore its load is maximized.* Only a small pressure shift is necessary. As a consequence, the VHP steam demand is reduced significantly.

Comparing the simple expansion plan (Fig. 7.6) with the alternative properly analyzed plan (Fig. 7.7), we see that there is significant reduction in both energy and capital costs as well as a corresponding reduction in emissions (see Table 7.1). The alternative plan *does not require investment in a new turbine or a new boiler.* It also does not require expansion of the cooling capacity. The alternative plan shows a reduction of 19 percent in the utilities operating cost and a global reduction of 15 percent in emissions (for which CO_2 is used as a proxy). In this particular instance, the percentage reduction in emissions is slightly less than that for utility costs, because the pinch solution reduces the amount of cogeneration on site. This results in an increase in electric import, and there is a corresponding increase in emissions at the external power station that supplies this electricity.

Reference 1 provides further options involving process and utilities modifications for the example case study.

By using the total site methodology, the most promising options for utilities and/or processes can be screened at the targeting stage. Since steam is simply an intermediate mechanism for heat transfer, its cost is no longer a factor. All designs are directly understandable in terms of fuel and power at the site boundaries.

The power afforded by the total site integration methodology has been a major breakthrough. Within the short time since its development it has gathered an impressive track record of successful applications.[2–6]

Summary: Total Site Targeting

Figure 7.8 summarizes the approach for overall site targeting.

1. Individual processes are represented through their grand composite curves. Economic decisions are made at this stage as to which process changes would be acceptable and which would not. Also, the initial trade-off is set in each process between process fuel and process steam.

2. Total site profiles are built up from the individual grand composite curves, and the site infrastructure is superimposed. Economic decisions are made again regarding which changes in the infrastructure would be acceptable.

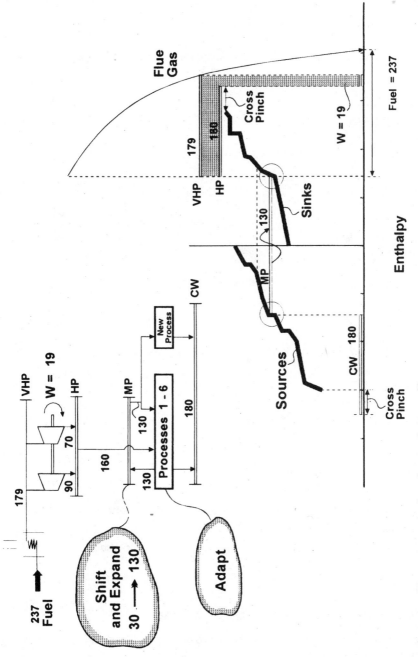

Figure 7.7 Alternative option based on total site profiles.

TABLE 7.1   Example Case Study: Comparison of Operating and Capital Costs

	Proposed design	Alternative design
Total utilities operating cost	100%	81%
CO_2 (global)	100%	85%
Capital cost implications	Expand MP steam New turbine New boiler Expand cooling	Shift MP steam and expand Adapt process to suit

Site power demand = 50 MW; cost data: fuel = $98/(kW · yr), power = $400/(kW · yr), cooling water = $5/(kW · yr).

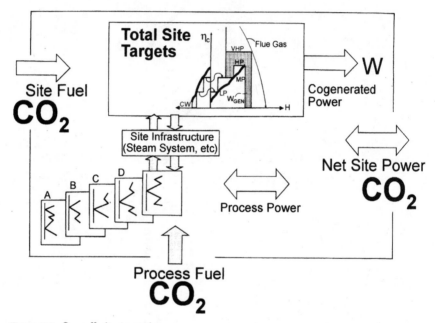

Figure 7.8   Overall site targeting.

3. Targets are set for site fuel, power, emissions, and cooling duties.

4. Decisions made earlier as to acceptable process and infrastructure changes are now tested for their importance in the overall context. If necessary, the procedure is iterated.

Road Map for Site Development

Once the site has been translated to total site profiles and different scenarios for site development have been explored, it is possible to establish relationships between investment and benefit, such as those

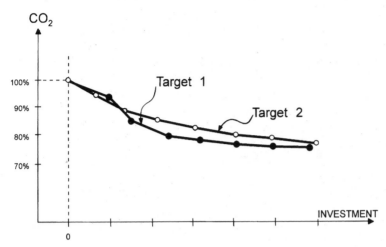

Figure 7.9 "Road map" for emissions targeting.

shown in Fig. 7.9.[7] In this case the desired "benefit" is reduction of CO_2 emissions; similar plots can be generated for energy savings as function of investment.

This representation is known as a *road map for site development*. Each curve (or route) relates to a set of assumptions made and a *project package*, i.e., a series of mutually compatible projects that can be implemented sequentially. Each project package is explored for its technical and economical feasibility. There will be (slight) differences in *value for money* (or emissions reduction for investment) between different packages. However, all packages examined will have been assessed against the targets set by more complicated schemes, involving more steam levels, etc., and will effectively present thermodynamic targets for acceptable economics.

The designer and/or planner can use the information from the "road map" to plan a "route" (strategy) for long-term site development. For example, if there is a requirement to reduce CO_2 emissions by, say, 10 percent over the next 5 years, the plots show the minimum investment needed to achieve this. The appropriate elements of the minimum cost project package can then be identified, detailed, and implemented within the required time.

This approach provides a high level of confidence that the planning scenario offers the most effective use of available capital. The targets established here are not based on statistics, learning curves, or "management by objectives." They are based on specific circumstances, conditions, and constraints on the specific site analyzed. They are based on technology and investment. They are rooted in thermodynamic analysis.

The road map described in this chapter relates primarily to the evaluation of heat integration options for reducing emissions within process sites. However, a similar methodology can be used for ranking virtually any alternative pollution prevention and pollution control technology on the basis of cost-effectiveness[8] (see Chap. 18).

References

1. V. R. Dhole and B. Linnhoff, "Total Site Targets for Fuel, Co-generation, Emissions, and Cooling," paper presented at ESCAPE-II Conference, Toulouse, France, October 1992. Also printed in *Comp. Chem. Eng.,* 17 (suppl.):s101–s109, 1993.
2. Y. Natori, "Managing the Implementation of Pinch Technology in a Large Company," paper presented at International Energy Agency Workshop on Process Integration, Gothenburg, Sweden, January 1992.
3. J. Snoek and T. N. Tjoe, "Process Integration Experience in a Large Company," paper presented at International Energy Agency Workshop on Process Integration, Gothenburg, Sweden, January 1992.
4. R. Davison, "Pinch Helps Energy Efficiency Drive," *Eur. Chem. News,* special report, July 1992.
5. R. J. Camm and D. N. Rihani, "Site Wide Optimization of a Refinery," paper presented at Chemtech Triple Expo '94, Bombay, India, November 1994.
6. A. P. Rossiter, M. A. Rutkowski, and V. R. Dhole, "Industrial Site Targets for Energy and Emissions," ASME Industrial Power Conference, Denver, CO, March 1993.
7. B. Linnhoff and V. R. Dhole, "Targeting for CO_2 Emissions for Total Sites," *Chem. Eng. Technol.,* 16:252–259, 1993.
8. A. P. Rossiter and J. D. Kumana, "Rank Pollution Prevention and Control Options," *Chem. Eng. Prog.,* February 1994, pp. 39–44.

Wastewater Minimization

Robin Smith, Ph.D.
Centre for Process Integration
University of Manchester Institute of Science and Technology
Manchester, England

Introduction

In the past, water has been assumed to be a limitless low-cost resource. However, attitudes are now changing. Public awareness of the danger to the environment caused by overextraction is now acute. The water industry has had to take note, and as a consequence, the price of fresh water for the process industries has escalated. At the same time, as part of the same ecological awareness, the imposition of ever-stricter discharge regulations has driven up waste treatment costs, which means capital expenditure with little or no productive return. There is now considerable incentive to reduce both freshwater consumption and wastewater production.

Water use on a typical process site is illustrated in Fig. 8.1.[1] Raw water is typically given some treatment, which might be as little as filtration. This water might then be used in various processes such as washing inorganics from products, vessel cleaning, and hosing operations. In these processes the water comes into contact with process materials, becomes contaminated, and is shown in Fig. 8.1 being sent to wastewater treatment. Fresh water is also upgraded in boiler feed-water (BFW) treatment for use in the steam system. Wastewater is

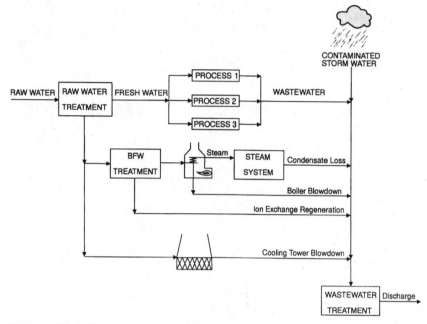

Figure 8.1 Typical water use on a chemical process site.

generated by ion-exchange regeneration, boiler blowdown, and condensate loss. Another major source of wastewater shown is cooling tower blowdown. These various wastewater streams are then typically mixed, along with contaminated storm water, and sent to treatment (Fig. 8.1). The wastewater is characterized by the volume and the contaminant load carried. The need is to reduce the consumption of fresh water and both wastewater volume and load with no detrimental effects on the processes.

Approaches to Water Minimization

There are four general approaches to water minimization:

1. *Process changes.* Process changes can reduce the inherent demand for water. For example, wet cooling towers can be changed to air coolers, or washing operations can have the number of stages increased.

2. *Water reuse.* Wastewater can be reused directly in other operations, provided the level of previous contamination does not interfere with the process, in arrangements such as that in Fig. 8.2a.[1] This reduces both freshwater and wastewater volume but leaves the load substantially unchanged.

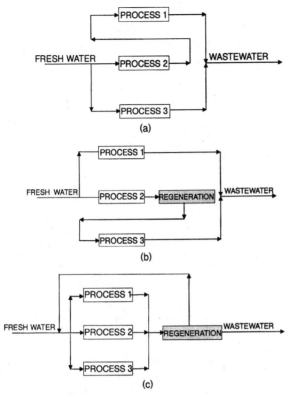

Figure 8.2   Water minimization through (*a*) reuse, (*b*) regeneration reuse, and (*c*) regeneration recycling.

3. *Regeneration reuse.*   Wastewater can be regenerated by partial treatment to remove the contaminants that would otherwise prevent reuse, and then it can be reused in other operations in arrangements such as that in Fig. 8.2*b*.[1] Regeneration is any operation that removes the contaminants which prevent reuse such as filtration, steam stripping, and carbon adsorption. It should be emphasized that when water is reused after regeneration, it does not reenter processes in which it has already been used. Regeneration reuse reduces freshwater and wastewater volume, and the load is reduced via the introduction of regeneration.

4. *Regeneration recycling.*   Wastewater can be regenerated to remove contaminants that have already built up, and then the water can be recycled in arrangements such as that in Fig. 8.2*c*.[1] In this case water can reenter processes in which it has already been used. Freshwater volume and wastewater volume are reduced, and again wastewater load is reduced via the introduction of regeneration.

It is important to distinguish between regeneration reuse and regeneration recycling, since recycling can create a buildup of undesired contaminants not removed in the regeneration.

To achieve these complex aims cost-effectively requires careful analysis and design.

Process Changes

The first instinctive measure taken to minimize water use is to make process changes that reduce the inherent demand. These could include, e.g.,

- Increasing the number of stages in extraction processes that use water
- Using spray balls for more effective internal vessel washing
- Changing from wet cooling towers to air coolers
- Improving control of cooling tower blowdown
- Fixing triggers to hoses to prevent unattended running
- Improving energy efficiency to reduce steam demand and hence reduce the wastewater generated by the steam system through boiler blowdown, aqueous waste from boiler feedwater treatment, and condensate loss
- Increasing condensate return from steam systems to reduce boiler blowdown and aqueous waste from boiler feedwater treatment
- Improving control of boiler blowdown

Such improvements tend to be made on an ad hoc basis with no consideration of the water-using processes as a system. Once a systematic approach is adopted, these changes are viewed from a different perspective. A systematic approach also directs changes to where they will be most effective. Such an approach will now be developed.

Targeting Minimum Wastewater for Single Contaminants

Thermodynamic pinch analysis is now a mature and effective tool for optimizing the minimum practical energy needs of a process and achieving these by using systematic design procedures[2,3] (see also Chap. 5). By taking advantage of certain parallels between the principles of heat and mass transfer, the targeting procedures of pinch analysis can be adapted to address water use and waste.[4-6]

Consider the mechanism by which process material is contacted with water in Fig. 8.3a. The water becomes contaminated as contaminant mass is transferred from the process material.

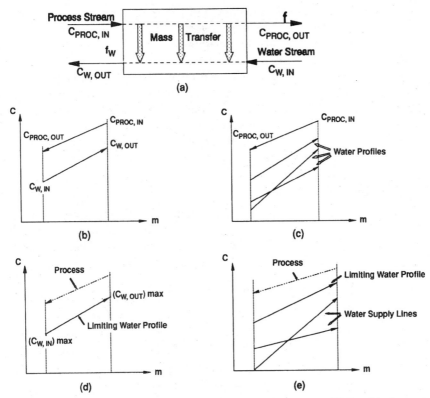

Figure 8.3 Representation of water use. (*a*) A water-using process. (*b*) Graphical representation of the water-using process. (*c*) Different water profiles can solve the problem. (*d*) Maximizing the inlet and outlet concentrations defines the limiting water profile. (*e*) Any water profile below the limiting water profile satisfies the process requirements. (From Wang and Smith, "Wastewater Minimization," *Chemical Engineering Science,* 49(7):981–1006. Copyright 1994. Reprinted with kind permission of Elsevier Science Ltd., The Boulevard, Langford Lane, Kidlington, OX 1GB, UK.)

Water-use representation

The process streams in Fig. 8.3*a* can be plotted as the concentration of contaminant versus the mass load of contaminant (see Fig. 8.3*b*). Inlet and outlet concentrations in the process stream, and hence the mass which is transferred, are fixed by the process specification. Different combinations of the flow rate and concentration in the water stream can satisfy the process requirements (Fig. 8.3*c*). The steeper the water-line in Fig. 8.3*c*, the smaller the flow rate. If the aim is to maximize the potential for reuse of water from other processes, the highest possible inlet concentration is needed. If the maximum possible outlet concentration is specified, then the water flow rate is minimized for the maximum inlet concentration (Fig. 8.3*d*). This water profile with maximum inlet and outlet concentration shown in Fig. 8.3d is not necessarily the profile that will be used in the final design. It simply represents a limiting case. It will therefore be designated the *limiting water profile*. Any *water supply line* below it can satisfy the process requirements (Fig. 8.3e). The use of the limiting water profile rather than the actual process stream data allows constraints due to concentration driving forces, solubility, precipitation, fouling, settling, and corrosion to be built into the analysis. This enables process operations in which the mass-transfer mechanisms are quite different to be considered on the same basis, e.g., a washing operation versus a steam stripper.

Limiting composite curves

Consider the process data given in Table 8.1.[5] When water is minimized for these processes by using a conventional approach, the result is shown in Fig. 8.4a. Fresh water is used in each of the four processes,

TABLE 8.1   Limiting Process Water Data

Process number	Contaminant mass load, kg/h	C_{in}, ppm	C_{out}, ppm	Water flow rate, t/h
1	2	0	100	20
2	5	50	100	100
3	30	50	800	40
4	4	400	800	10

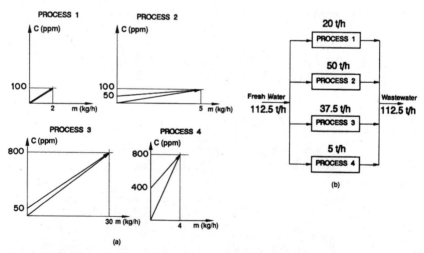

(a)

(b)

Figure 8.4 A conventional approach to water minimization for the data of Table 8.1. (a) Minimizing fresh water use in each individual process. (b) Design for minimum water flow rates which uses fresh water in each process. (From Wang and Smith, "Wastewater Minimization," *Chemical Engineering Science*, 49(7):981–1006. Copyright 1994. Reprinted with kind permission of Elsevier Science Ltd., The Boulevard, Langford Lane, Kidlington, OX 1GB, UK.)

and its flow rate minimized by maximizing the outlet concentration to the values given in Table 8.1. The resulting design is shown in Fig. 8.4*b* and requires a total of 112.5 t/h of fresh water, generating the same amount of wastewater.

Note that transfer is assumed to be a linear function of concentration. This is valid for dilute streams. For processes with significantly nonlinear behavior, the stream can be broken down into a series of linear segments.

To minimize the overall water flow rate, there must be an overall analysis of the water-using processes. Figure 8.5*a* shows the process data from Table 8.1 plotted on the same axes. The inlet and outlet concentrations of each process stream define concentration intervals within which the rate of change of mass with concentration is constant. In Fig. 8.5*b* the mass loads of the individual streams are combined within each interval to produce a *limiting composite curve* (Fig. 8.5*b*). This represents how the processes would behave if they were a single process rather than four separate processes.

If the incoming fresh water has no contamination, a water supply line can be plotted from zero to be matched to the limiting composite curve. The steeper the slope of the water supply line, the greater the concentration change for a given amount of contaminant picked up and the lower the flow rate. Figure 8.5*c* shows a water supply line with the steepest slope and maximum outlet concentration. The point of contact between the limiting composite curve and the water supply line which limits the slope is the *pinch*. At the pinch, mass-transfer driving forces have gone to a minimum or concentrations have gone to limiting values set by corrosion limitations, etc. The target value for the minimum

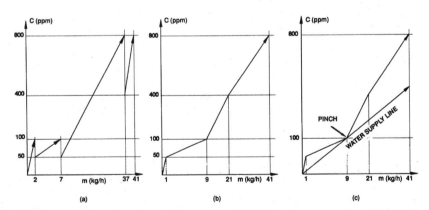

Figure 8.5   The limiting composite curve. (*a*) The limiting water profiles. (*b*) Limiting composite curve. (*c*) Matching the water supply line defines the minimum water flow rate. (From Wang and Smith, "Wastewater Minimization," *Chemical Engineering Science*, 49(7):981–1006. Copyright 1994. Reprinted with kind permission of Elsevier Science Ltd., The Boulevard, Langford Lane, Kidlington, OX 1GB, UK.)

water flow rate for this problem is 90 t/h, which is 20 percent less than the original 112.5 t/h in Fig. 8.4. For this target of 90 t/h to be meaningful, it must be possible to design a network in which concentrations do not exceed the maximum permitted for inlet and outlet and in which the same mass load of contaminant is picked up.

Design for Minimum Wastewater for Single Contaminants

Two approaches to design are possible: the maximum driving forces and the minimum number of water sources.[5] These have different objectives, but both are able to achieve the minimum flow rate target within the permitted concentration constraints. The maximum driving forces approach often leads to designs that have some unnecessary complexity. It is therefore not considered further here.

In most cases it is desirable in water networks to use a single source of water for each process, which means using a single match for each process in the network design while still achieving the target. This is the objective of the minimum number of water sources approach.[5]

In designing heat exchanger networks, it is known that bypassing and mixing can be used to reduce the number of units below the apparent minimum for problems where a pinch divides the problem into two parts.[5] The technique is not normally used in heat-exchanger networks since it tends to lead to a greatly increased heat-transfer area. However, in the case of water networks, the technique of bypassing and mixing can be used much more effectively.

The approach now is to use concentration intervals in which the driving forces are minimized in individual matches, only sufficient water being used to maintain feasibility and any excess being bypassed for use later. The limiting composite curve is now divided horizontally into concentration intervals (see Fig. 8.6a). The grid diagram[7,8] is then used to develop the water network in these concentration intervals (Fig. 8.6b). Again, starting in interval 1, only process 1 exists and is matched with the fresh water. However, only as much water as is necessary to maintain feasibility, that is, 20 t/h giving a concentration of 50 ppm at the outlet, is used with the remainder being bypassed for use later. This degrades the minimum volume of water in interval 1. In interval 2, processes 1, 2, and 3 exist together, and water is split between them to obtain 100 ppm at their outlet. In interval 3 there is only process 3, so the minimum outlet water from interval 2 is used and the remainder is bypassed. Finally, in interval 4 both processes 3 and 4 exist. The water is split between these, and any excess above the minimum is bypassed. The resulting flow sheet is shown in Fig. 8.6c. The target (90 t/h) is achieved.

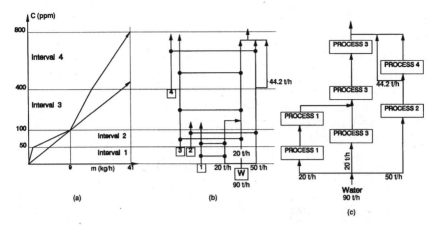

Figure 8.6 Design for minimum number of water sources. (*a*) Limiting composite curve. (*b*) Design grid. (*c*) Conventional flow sheet. (From Wang and Smith, "Wastewater Minimization," *Chemical Engineering Science,* 49(7):981–1006. Copyright 1994. Reprinted with kind permission of Elsevier Science Ltd., The Boulevard, Langford Lane, Kidlington, OX 1GB, UK.)

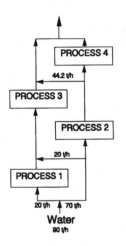

Figure 8.7 Simplification of the design for minimum number of water sources. (From Wang and Smith, "Wastewater Minimization," *Chemical Engineering Science,* 49(7):981–1006. Copyright 1994. Reprinted with kind permission of Elsevier Science Ltd., The Boulevard, Langford Lane, Kidlington, OX 1GB, UK.)

Although it meets the necessary criteria for the minimum flow rate, the design is unnecessarily complex. Linnhoff and Hindmarsh introduced a method of energy relaxation for the simplification of heat-exchanger networks.[8] A similar method for simplification can be applied to water networks such as that in Fig. 8.6. Figure 8.7 shows the final design as a conventional flow sheet. This design now features a single source of water for each process. Further evolution is possible.

Multiple Contaminants

Truly single-contaminant cases occur rarely. However, the approach developed so far for single contaminants can often be applied directly to multiple-contaminant situations by representing multiple-contaminant behavior using pseudo-single-contaminant aggregate properties such as suspended solids, total dissolved solids, and chemical oxygen demand and then analyzing as a single contaminant. Nevertheless, many cases will still call for a genuine multiple-contaminant approach; and the mass-transfer pinch methodology can be extended to deal with these situations.[5]

Wastewater Regeneration

So far it has been assumed that the water is reused without treatment. However, regeneration can be used to remove contaminants on an intermediate basis, by processes such as gravity settling, filtration, and membranes. These treatment processes can be used singly or in combination. The choice and placement of regenerators are crucial. In addition, it is important to distinguish between regeneration reuse and regeneration recycling.

Figure 8.8a shows the construction of the water supply line against the limiting composite curve with the inclusion of a regeneration process which reduces the level of contamination to C_0. The inlet concentration to the regeneration process in Fig. 8.8a is the pinch concentration C_{pinch}. At first sight this seems to be infeasible since the water supply line with its regeneration crosses the limiting composite curve. However, to deter-

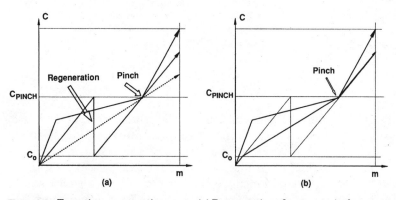

Figure 8.8 Targeting regeneration reuse. (a) Regeneration of water at pinch concentration. (b) A composite of the water supply line before and after regeneration indicates the water flow rate minimized. (From Wang and Smith, "Wastewater Minimization," *Chemical Engineering Science,* 49(7):981–1006. Copyright 1994. Reprinted with kind permission of Elsevier Science Ltd., The Boulevard, Langford Lane, Kidlington, OX 1GB, UK.)

mine the overall feasibility, a composite of the water supply line before
and after the regeneration must be constructed as shown in Fig. 8.8b.
This is now seen to be just feasible. Allowing the water to reach the pinch
concentration before regeneration achieves two criteria simultaneously
for regeneration reuse. The minimum water flow rate is achieved together with the minimum concentration reduction in the regeneration, compatible with that minimum water flow rate.[5]

Note that in Fig. 8.8a the slope of the water supply line before and
after regeneration is the same. This means that there is no change in
the flow rate through the regeneration process. For some regeneration
processes this would be a good approximation; for others it would not.
However, the construction is easily modified to allow for changes in
flow rate by relating the slopes before and after regeneration to the
flow rate change characteristics of the regeneration process.

If recycling of water is allowed, then even less fresh water is needed
and the minimum flow rate is further reduced. Figure 8.9a shows a
water supply line with regeneration in which recycling is allowed. The
incoming fresh water is reduced to a minimum flow rate which is set by
the slope of the limiting composite curve below C_0. After regeneration
the water flow rate increases because of recycled water which has been
regenerated. Figure 8.9b shows a composite of the water supply line
which is seen to be feasible.

The approach also allows the specification of regeneration processes
by using the removal ratio rather than a specified outlet concentration.
Cases which call for partial regeneration are also identified. Targeting
for regeneration reuse and regeneration recycling can also be extended
to multiple contaminants.[5]

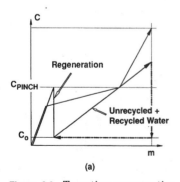
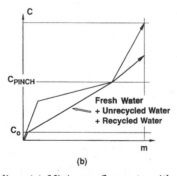

(a)   (b)

Figure 8.9 Targeting regeneration recycling. (a) Minimum flow rate with
regeneration recycling. (b) A composite of the water supply before and after
regeneration. (From Wang and Smith, "Wastewater Minimization," *Chemical
Engineering Science*, 49(7):981–1006. Copyright 1994. Reprinted with kind
permission of Elsevier Science Ltd., The Boulevard, Langford Lane,
Kidlington, OX 1GB, UK.)

By using the data given in Table 8.1, assuming a regeneration process capable of removing contaminants down to a concentration of 5 ppm, and applying these rules, regeneration reuse produces a reduced flow rate of 46.2 t/h with the design shown in Fig. 8.10. Regeneration recycling further reduces the value to 20 t/h for the data in Table 8.1, with the design in Fig. 8.11.[5]

These conceptual insights make it easy to specify the most appropriate types of regenerator, to determine their placement for any combi-

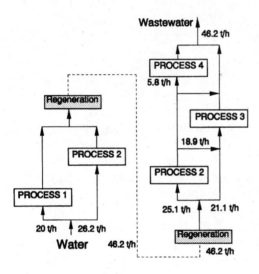

Figure 8.10 Design with regeneration reuse. (From Wang and Smith, "Wastewater Minimization," *Chemical Engineering Science,* 49(7):981–1006. Copyright 1994. Reprinted with kind permission of Elsevier Science Ltd., The Boulevard, Langford Lane, Kidlington, OX 1GB, UK.)

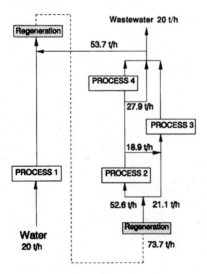

Figure 8.11 Design with regeneration recycling. (From Wang and Smith, "Wastewater Minimization," *Chemical Engineering Science,* 49(7):981–1006. Copyright 1994. Reprinted with kind permission of Elsevier Science Ltd., The Boulevard, Langford Lane, Kidlington, OX 1GB, UK.)

nation of single and multiple contaminants, and to distinguish between reuse and recycling, for the minimum overall water flow rate. Simple representations are produced, with targets that allow flow sheet scenarios to be screened without resorting to repeated design.

Case Study

Petroleum refining consumes very large quantities of water and generates a large volume of wastewater. Here three refinery processes are considered: a distillation using live steam injection, a hydrodesulfurization process, and a desalter. The last two processes use water to wash out contaminants. Table 8.2 gives the data for three such process streams. It is necessary in this case to consider three contaminants: oil, H_2S, and salt.[5,9]

Recycling is not permitted because of inorganic buildup, but regeneration is possible with a foul-water stripper having removal ratios of 0.0, 0.999, and 0.0, respectively, on hydrocarbon, H_2S, and salt. Cost correlations are given in Table 8.3.[9]

Applying the procedures produces the target profiles shown in Fig. 8.12, which are realized in the evolved minimum sources design with

TABLE 8.2   Process Data for Refinery Case Study

Process number	Flow rate, t/h	Contaminants	C_{in}, ppm	C_{out}, ppm
1	45	Oil	0	15
		H_2S	0	400
		Salt	0	35
2	34	Oil	20	120
		H_2S	300	12,500
		Salt	45	180
3	56	Oil	120	220
		H_2S	20	45
		Salt	200	9,500

TABLE 8.3   Economic Data for Refinery Case Study

Annual rate of return,%	10
Operating hours, h/yr	8600
Freshwater cost $/t	0.30
Cost of end-of-pipe treatment	
Investment, $	$34{,}200f^{0.7}$
Operating, $/h	$1.0067f^{0.7}$
Cost of foul-water stripper	
Investment, $	$16{,}800f^{0.7}$
Operating, $/h	$1.0f^{0.7}$

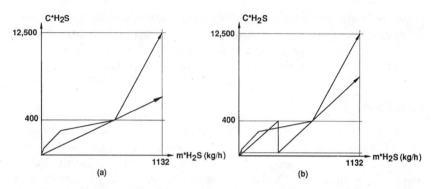

Figure 8.12 Targeting for the refinery case study. (a) Target without regeneration ($f =$ 105.7 t/h). (b) Target for regeneration reuse ($f =$ 55.5 t/h). (From Wang and Smith, "Wastewater Minimization," *Chemical Engineering Science,* 49(7):981–1006. Copyright 1994. Reprinted with kind permission of Elsevier Science Ltd., The Boulevard, Langford Lane, Kidlington, OX 1GB, UK.)

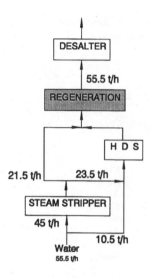

Figure 8.13 Design for the refining case study with regeneration reuse. (From Wang and Smith, "Wastewater Minimization," *Chemical Engineering Science,* 49(7):981–1006. Copyright 1994. Reprinted with kind permission of Elsevier Science Ltd., The Boulevard, Langford Lane, Kidlington, OX 1GB, UK.)

regeneration shown in Fig. 8.13. A comparison of results is given in Table 8.4, which demonstrates cost reductions of 20 percent for reuse alone and 26 percent for regeneration reuse.

Overall Strategy for Water Minimization

Earlier, the traditional approaches to water minimization such as changing washing operations and improving control of boiler blowdown were discussed. Although these measures obviously have a role to play

TABLE 8.4   Comparison of Results

	Flow rate, t/h	Annual cost, million $
Without reuse	133	1.599
Reuse without regeneration	106	1.281
Reduction		
Δ	27	0.318
Percent	20	20
Regeneration reuse	55.5	1.186
Reduction		
Δ	77.5	0.413
Percent	58	26

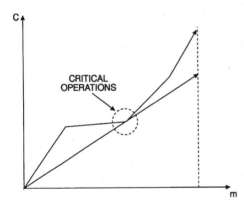

Figure 8.14 Targets point to where process changes and good housekeeping will be most effective.

in water minimization, the approach presented here both complements them and allows them to be directed to where they will be most effective. Figure 8.14 shows a limiting composite curve with the water supply line matched to it. The critical operations, which limit the extent to which water can be minimized in an overall sense, are those around the pinch. It is to these processes that process changes should be directed. Away from the pinch, process changes have no influence on overall water use and wastewater generation.

Although comparatively new, the methodology has been successfully tested on several industrial sites. Applications have ranged from petroleum refining to specialty batch chemical operations, with water savings ranging from 30 to 60 percent.[6]

The techniques described in this chapter allow targeting for maximum water reuse, minimizing both fresh water and wastewater. The appropriate placement and size of the regenerator, with either reuse or recycling, are targeted, taking into account constraints due to driving forces, equipment fouling, corrosion limitations, and the presence of multiple contaminants. Such targets allow the designer to quickly scope and screen alternative scenarios without actually having to carry out the design. Once acceptable targets have been achieved, two alternative design procedures allow the targets to be achieved in practice. The first design method makes best use of concentration driving forces while the second allows design with a single source of water in each process. This latter method is the one most often used in practice.

Sometimes, processes require a fixed flow rate rather than one which is allowed to vary with the inlet concentration. For example, vessel cleaning, hosing operations, etc., tend to require a fixed flow rate irrespective of the concentration of contamination at the inlet. An additional complication is that in some processes there is a fixed flow rate which is lost and cannot be reused. For example, the makeup to cooling towers to allow for evaporation is a fixed flow rate which is lost. Yet another complication arises when there are several sources of fresh water available with different qualities. All these features can be included in the analysis.[10]

References

1. B. Linnhoff and R. Smith, "Recent Developments in Pinch Analysis," paper presented at Process Systems Engineering 1994, Kyongju, Korea, 1994.
2. B. Linnhoff, D. W. Townsend, D. Boland, G. F. Hewitt, B. E. A. Thomas, A. R. Guy, and R. H. Marsland, "User Guide on Process Integration for the Efficient Use of Energy," *IChemE*, Rugby, UK, 1982.
3. R. Smith, *Chemical Process Design*, McGraw-Hill, New York, 1995.
4. M. M. El-Halwagi and V. Manousiouthakis, "Synthesis of Mass-Exchange Networks," *AIChE Journal*, 8:1233–1244, 1989.
5. Y. P. Wang and R. Smith, "Wastewater Minimization," *Chem. Eng. Sci.*, 49(7):981–1006, 1994.
6. R. Smith, E. Petela, and Y. P. Wang, "Water, Water Everywhere," *The Chem. Eng.*, no. 565, pp. 21–24, May 12, 1994.
7. B. Linnhoff and J. R. Flower, "Synthesis of Heat Exchanger Networks. Part I: Systematic Generation of Energy Optimal Networks," *AIChE Journal*, 24(4):633–642, 1978; "Part II: Evolutionary Generation of Networks with Various Criteria of Optimality," *AIChE Journal*, 24(4): 642–654, 1978.
8. B. Linnhoff and E. Hindmarsh, "The Pinch Design Method of Heat Exchanger Networks," *Chem. Eng. Sci.*, 38(5):745–763, 1983.
9. N. Takama, T. Kuriyama, K. Shiroko, and T. Umeda, "Optimal Water Allocation in a Petroleum Refinery," *Comp. Chem. Eng.*, 4:251–258, 1980.
10. Y. P. Wang and R. Smith, "Wastewater Minimization with Flowrate Constraints," *Trans. IChE*, part A, accepted for publication.

Knowledge-Based Approaches

9

Introduction to Knowledge-Based Approaches

Alan P. Rossiter, Ph.D.
Linnhoff March, Inc.
Houston, Texas

Introduction

Many features are common to virtually all processes across a wide range of industrial sectors. Recognition of this fact has led to the development of several procedures for process synthesis and process integration that can be categorized as *knowledge-based approaches*. The procedures vary considerably in detail and sophistication, ranging from simple lists of heuristics (rules of thumb), through databases and hierarchical design procedures, to artificial intelligence programs. Each of these approaches has proved useful in numerous applications, and all build upon an accumulated *knowledge base* of proven ideas to develop good processes or process improvements with the minimum of engineering effort.

The same concepts that apply to process design in general are also applicable to the specific problems of designing and revamping plants to reduce pollution. This has led to the introduction of a number of knowledge-based approaches to waste minimization. These draw on an understanding of the universal features of industrial processes to identify improved emissions reduction opportunities more easily and more rapidly than would otherwise be possible.

Universal Features of Industrial Processes

The universal features of industrial processes include the processing steps for material conversion, material separations, material recycling, and energy utilization, as well as infrastructure issues, notably storage of feed and product materials.

Typically, processes involve the conversion of one or more feeds into one or more products. This requires some type of material conversion step, commonly a reactor, but in some cases other unit operations (such as crystallization or digestion) may be involved.

The material from the conversion step is usually not the final product: Unconverted feed materials and by-products or wastes have to be removed. Thus some type of separation is needed. Once again, the type of separation system varies with the process: Distillation is generally regarded as the workhorse of the refining and petrochemicals industries; fractional crystallization is often used for close-boiling systems; various types of solid-liquid separators (filters, centrifuges, and hydrocyclones, to mention but a few) are common in processes making crystalline inorganic and organic products. Paper machines (separating pulp from white water) are a specific example of solid-liquid separation in the pulp and paper industry. However, notwithstanding the array of equipment types, the basic requirement for processes is basically the same across all the industrial sectors: separation of material into product and nonproduct streams.

The fate of the separated materials is also similar in most processes. Streams intended as products may be in their final form as they leave the separation system, or they may require some further processing step or steps (such as drying, agglomerating, or further purification). Nonproduct streams are generally recycled or purged. Thus there are only a few structural options open in the design of most processes.

Energy is another common denominator for all industrial processes. Thermal energy is typically supplied by steam, fuel-fired heaters, or electric heaters of various types. Heat may be removed by ambient cooling (air or water) or at lower temperatures by refrigeration. Heat may also be recovered within a process by heat exchange between process streams. Mechanical energy (usually derived from steam turbines, gas turbines, or electricity) is used for pumps and compressors and to drive other items of process equipment.

Maintaining inventories (of both raw materials and product) creates a need for storage and shipping facilities. These can vary considerably, in both size and form, depending on the physical state and toxicity of the material and the volume produced. They range from large open piles of wood in the forest products sector to high-pressure storage spheres for LPG. Nevertheless these facilities that are very different physically all have very similar requirements for inventory control.

The differences between individual processes are obviously significant and should not be ignored. However, the foregoing discussion illustrates an important truth: The similarities are far greater than the differences, even for industries as varied as oil refining, pulp and paper, pharmaceuticals manufacture, and food processing. This premise forms the foundation of the knowledge-based methodologies that have been developed for generating new process designs and identifying promising retrofit options.

Some of these procedures have been adapted in recent years to provide a more specific focus on waste minimization and pollution prevention. These include

- Methods designed to transfer pollution prevention solutions directly from one application to others (Chaps. 10 and 11)

- Hierarchical design and review (Chap. 12)

- Artificial intelligence (Chaps. 13 and 14)

This chapter provides a general discussion of these methodologies.

Heuristics

An idea that has been proved in one application is very likely to be useful in others, too. This premise is firmly embedded in the theory and practice of process design, and texts on the subject[1] are littered with examples. These are often formalized as *heuristics*.

Heuristics, or rules of thumb, provide guidelines for the development of good designs or retrofit options. They are typically based on accumulated experience from a large number of similar applications, and they represent the distilled wisdom of many engineers. Where heuristics are available, it is often possible to go a long way toward defining good design options with only limited calculation and experimentation, thereby saving much work. If heuristics are not available, the burden of calculation and test work is greater. For example, the statement "...the minimum on the total-cost curve will generally occur at an operating reflux ratio of 1.1 to 1.5 times the minimum R..."[2] constitutes a heuristic for optimizing distillation columns. Armed with this, the engineer can often develop a near-optimal distillation design without carrying out rigorous optimization computations.

It is important to understand the limitations of heuristics. The above distillation heuristic, e.g., is applicable only to *stand-alone* columns. Very different optimum reflux ratios will be obtained with heat-integrated columns. Moreover, heuristics do not provide definitive designs. Indeed, it is not uncommon for different heuristics to yield conflicting conclusions. For example, several heuristics have been proposed for sequencing distillation columns for separating multicomponent mixtures.[3] These include recommendations such as "remove most abundant component first," "remove corrosive materials first," and "products and recycles should be taken as overheads." Clearly, there are circumstances in which these heuristics will be in conflict, yielding different preferred sequences.

This example typifies an underlying problem with the use of heuristics: They are simply guidelines based on a body of experience, and they cannot guarantee a "right" answer in every case. They do, nevertheless, provide helpful insights and can be used for rapid generation of ideas for evaluation.

Heuristics, in one form or another, provide the foundation for the knowledge-based approaches to process design. They may be combined with a systematic method of analyzing flow sheets or with databases; they may be computerized or exist only as hard copy; they may be formalized as generalizations or simply presented as informal statements of "ideas that work." The key, in each of these approaches, is that we can build on the accumulated experience of many engineers by using heuristics.

Transferring Pollution Prevention Solutions

A number of successful pollution prevention solutions have been published.[4,5] Even without formalizing ideas like these explicitly as heuristics, it is often possible to adapt them for use in new situations. If an engineer is confronted with a specific problem, there may be a clear analogy with a published solution. In such cases the solution can be transferred directly to the new application. However, even when no direct comparison exists, published solutions can provide a basis for discussion and brainstorming, leading ultimately to the generation of implementable design options for process improvements. Chapter 10 summarizes a number of such ideas, based on practical plant experience.

Chapter 11 introduces a rather more structured approach for transferring ideas to new applications, based on a *unit operations approach*.[6] This uses a database of waste minimization solutions, classified according to the type of unit operation in which the solution is applied. This information, together with a set of ranking parameters, is used to identify options for addressing any given waste minimization problem and to provide an initial screening of the ideas that are identified.

Hierarchical Decision and Review

Hierarchical review has its origins in the hierarchical decision approach to process design.[1] The basic premise is that process designs evolve via a series of decisions, which provide successive levels of detail. By identifying the key decisions and implementing them in the correct order, it is possible to identify good design options with minimal effort and rework. Heuristics, combined with simple calculations, are used to generate alternative design options at each decision level. An economic evaluation of the process can be carried out at each level, providing a screening tool to eliminate nonviable processes without having to expend excessive design effort.

The hierarchical approach was first applied to the design of petrochemical processes. It has since been used in the design of a variety of different types of plant, such as solids processes,[7] batch processes,[8] and polymers.[9] A similar overall hierarchical structure can be developed for each class of problem, although the detailed questions that must be addressed, especially at some of the lower decision levels, depend on the type of problem being analyzed.

The method has also been adapted for retrofit applications. In this case, rather than providing a logical sequence for making design decisions, the procedure provides a systematic framework for reviewing and improving existing installations. An early publication on the use of this approach for retrofits related to an Imperial Chemical Industries solids process.[10] Significantly, this study identified a waste minimization opportunity, even though this was not the specific focus of the work. More recently, in work reported by Union Carbide, the hierarchical concept was combined with numerical optimization procedures in a more sophisticated retrofit procedure.[11]

In the past, the focus of hierarchical studies has been on optimizing the trade-offs between capital and operating costs. In the present context, the emphasis is on reduction of emissions, so the focus changes: The objective becomes the generation of new design or retrofit options for source reduction or beneficial recycling.[12,13] Where appropriate, end-of-pipe options are also identified. The details of the procedure are discussed in Chap. 12.

Note that the procedure cannot guarantee "optimum" designs, nor does it guarantee "minimum achievable" emissions. What it does provide is an objective framework for generating, screening, and comparing options and defining realistically attainable targets for emission levels by using available technologies. Given this approach, improved solutions can invariably be found more quickly than with conventional methods.

Artificial Intelligence

Artificial intelligence (AI) is an "extensive branch of computer science embracing pattern recognition, knowledge-based systems, computer vision, robotics, scene analysis, natural language processing (and) mechanical theorem proving, with more areas being added all the time."[14] It is based on the "concept that computers can be programmed to assume capabilities thought to be like human intelligence such as learning, reasoning, adaptation and self-correction."[15]

Process synthesis and design problems have elements that are not convergent mathematical problems, with a single correct solution. They involve divergence of thought and logical as well as mathematical analysis. This makes them good candidates for the application of artificial intelligence. For example, Douglas' hierarchical design procedure has been used as the basis for an expert system.[16] Other approaches to process synthesis and design have also been incorporated into AI systems, with the specific focus of waste minimization.[17,18] In general, these rely on the generation of a systematic synthesis procedure, together with heuristic rules for implementing waste minimization strategies. Chapter 13 provides a more complete introduction to this subject, with two short illustrative examples. Chapter 14 presents a more detailed case study relating to refinery waste minimization.

At the time of this writing, AI has not yet received widespread industrial acceptance in process synthesis and design applications. However, in view of its successes in other engineering fields and the growing body of academic experience, it seems highly likely that over the next few years AI will become an important tool for aiding process design for waste minimization.

References

1. J. M. Douglas, "A Hierarchical Decision Procedure for Process Synthesis," *AIChEJ*, 31(3):353–362, March 1985.
2. R. H. Perry and D. Green (eds.), *Perry's Chemical Engineers' Handbook*, 6th ed., McGraw-Hill, New York, 1984, pp. 13–34.
3. J. M. Douglas, *Conceptual Design of Chemical Processes*, McGraw-Hill, New York, 1988, p. 177.
4. K. E. Nelson, "Use These Ideas to Cut Waste," *Hydroc. Proc.*, 69(3):93–98, March 1990.
5. D. M. Haseltine, "Wastes: To Burn, or Not to Burn?" *Chem. Eng. Progress*, 88(7):53–58, July 1992.
6. S. A. Slater, J. Khan, and R. L. Smith, "Transferring Waste Minimization Solutions between Industrial Categories with a Unit Operations Approach: I. Chemical and Plating Industries," American Chemical Society I&EC Special Symposium on Emerging Technologies for Hazardous Waste Management, Atlanta, GA, September 21–23, 1992.
7. A. P. Rossiter and J. M. Douglas, "Design and Optimization of Solids Processes Part 1: A Hierarchical Decision Procedure for Process Synthesis of Solids Systems," *Chem. Eng. Res. Des.*, 64:175–183, May 1986.
8. M. F. Malone and O. Iribarren, "A Systematic Procedure for Batch Process Synthesis," AIChE Annual Meeting, Chicago, November 1985.
9. T. F. McKenna, "The Conceptual Design of Polymer Producing Processes," Ph.D. thesis, University of Massachusetts, September 1989.
10. A. P. Rossiter, D. C. Woodcock, and J. M. Douglas, "A Hierarchical Decision Procedure for Retrofit Studies of Solids Systems," American Institute of Chemical Engineers national meeting, paper 38b, Miami Beach, November 1986.

11. N. M. Duty, "A Random Search Approach to Optimized Process Retrofits," American Institute of Chemical Engineers spring national meeting, Houston, April 1991.
12. J. M. Douglas, "Process Synthesis for Waste Minimization," *Ind. Eng. Chem. Res.*, 31:238–243, 1992.
13. A. P. Rossiter, H. D. Spriggs, and H. Klee, Jr., "Apply Process Integration to Waste Minimization," *Chem. Eng. Progress*, 89(1):30–36, January 1993.
14. P. M. B. Walker (ed.), *Chambers Science and Technology Dictionary*, W & R Chambers and The Press Syndicate of the University of Cambridge, 1988.
15. Ibid.
16. R. L. Kirkwood, M. H. Locke, and J. M. Douglas, "A Prototype Expert System for Synthesizing Chemical Process Flowsheets," *Comput. Chem. Eng.*, 12:329–343, 1988.
17. T. F. Edgar and Y. L. Huang, "An Artificial Intelligence Approach to the Design of a Process for Waste Minimization," American Chemical Society I&EC Special Symposium on Emerging Technologies for Hazardous Waste Management, Atlanta, GA, September 21–23, 1992.
18. Y. L. Huang and L. T. Fan, "Artificial Intelligence for Waste Minimization in the Process Industries," *Comp. Ind.*, 22:117–128, 1993.

10

Process Modifications That Save Energy, Improve Yields, and Reduce Waste*

Kenneth E. Nelson

President
KENTEC Incorporated
Baton Rouge, Louisiana

Introduction

The future of the chemical industry is closely tied to conserving energy, improving yields, and reducing waste. The author has been closely associated with these activities for many years and has a wealth of experience in this important field. This chapter is designed to share some of that information.

What Is Waste?

Let's start by defining waste. Waste includes streams or materials that are

- Vented to the air
- Discharged to the water

---

*Copyright Dow Chemical Company.

- Sent to landfill
- Sent to an incinerator
- Sent to a flare
- Sent to a biological treatment facility

Waste reduction can be accomplished by creating less waste initially or by recycling waste products back to the process. It is also possible to reduce emissions by operating waste treatment units more efficiently, but that is not the concern of this chapter. Also, we will not discuss the nature of the waste (e.g., hazardous or nonhazardous, toxic or nontoxic, flammable or nonflammable).

This chapter is not meant to be totally comprehensive. The author has not made an exhaustive study of all possible ways of reducing waste. But it does contain a wealth of practical ideas and is intended to serve as the basis for discussion and brainstorming. We will begin with ideas associated with raw materials.

Raw Materials

Raw materials are usually purchased from an outside source or are transferred from an on-site plant. Each raw material needs to be studied to determine how it affects the amount of waste produced. The specifications for each raw material entering the plant should be closely examined.

1. *Improve quality of feeds.* Although the percentage of undesirable impurities in a feed stream may be low, it can be a major contributor to the total waste produced by a plant. Reducing the level of impurities may involve working with the supplier of a purchased raw material, working with on-site plants that supply feed streams, or installing new purification equipment. Sometimes the effects are indirect (e.g., water gradually kills the reactor catalyst, causing formation of by-products, so a drying bed or column is added).

2. *Use off-specification material.* Occasionally, a process can use off-specification material (that would otherwise be burned or sent to a landfill) because the particular quality that causes the material to be out of specification is not important to the process.

3. *Improve the quality of products.* Impurities in your own products may be creating waste in your customers' plants. Not only may this be costly, but also it may cause customers to look elsewhere for higher-quality raw materials. Take the initiative in discussing the effects of impurities with your customers.

4. *Use inhibitors.* Inhibitors prevent unwanted side reactions or polymer formation. A wide variety of inhibitors are commercially available. If inhibitors are already being used, check with suppliers for improved formulations and new products.

5. *Change shipping containers.* If raw materials are being received in containers that cannot be reused and need to be burned or sent to a landfill, change to reusable containers or bulk shipments. Similarly, using alternative shipping containers for sending products to customers should be considered.

6. *Reexamine the need for each raw material.* Sometimes the need for a particular raw material (one which ultimately ends up as waste) can be reduced or eliminated by modifying the process. The need for algae inhibitors in a cross-flow cooling tower, e.g., has been cut in half by shielding the water distribution decks from sunlight.

Reactors

The reactor is the heart of the process, and it can be a primary source of waste products. The quality of mixing in a reactor is crucial. Too often, insufficient design time is spent on scaling up this key parameter, and a new production facility has disappointing yields compared to laboratory or pilot plant data. Several of the ideas listed below are concerned with the quality of mixing.

1. *Improve physical mixing in the reactor.* Modifications to the reactor such as adding or improving baffles, installing a higher-rpm motor on the agitator(s), or using a different mixer blade design (or multiple impellers) can improve mixing. Pumped recirculation can be added or increased. Two fluids going through a pump, however, do not necessarily mix well, and a static mixer may be needed to ensure good contacting.

2. *Distribute feeds better.* Here is an area that deserves more attention than it typically gets. The problem is illustrated in Fig. 10.1. Reactants entering at the top of a fixed catalyst bed are poorly distributed. Part of the feed short-circuits down through the center of the reactor so that there is inadequate time for conversion to the desired products. The feed closer to the walls remains in the reactor for too long a time and "overreacts" to by-products that eventually become waste. A solution is to add some sort of distributor that causes the feed to move uniformly through all parts of the reactor as shown on the right. Some sort of special collector at the bottom of the reactor may also be necessary to prevent the flow from necking down to the outlet. Similarly, if a gas reactant is added to a liquid, the gas needs to be finely dispersed and evenly distributed throughout the liquid phase.

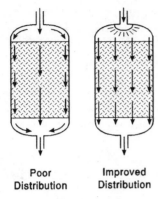

Poor   Improved
Distribution   Distribution

Figure 10.1 Types of flow distribution in reactors.

3. *Improve catalyst.* Searching for better catalysts should be an
ongoing activity because of the significant effect a catalyst has on the
reactor conversion and product mix. Changes in the chemical makeup
of a catalyst, the method by which it is prepared, or its physical characteristics (size, shape, porosity, etc.) can lead to substantial improvements in catalyst life and effectiveness.

4. *Improve the way in which reactants are added.* The idea here is
to get closer to ideal reactant concentrations before the feed enters the
reactor. This helps avoid secondary reactions which form unwanted by-
products. The way not to add reactants is shown in the upper half of
Fig. 10.2. It is doubtful that the "ideal" concentration exists anywhere
in this reactor. A consumable catalyst should be diluted in one of the
feed streams (one which does not react in the presence of the catalyst).
The lower half of Fig. 10.2 illustrates one approach to improving the
situation by using three in-line static mixers.

5. *Examine heating and/or cooling techniques.* The technique for
heating or cooling the reactor needs to be examined, especially to avoid
hot spots or overheated feed streams, both of which usually give
unwanted by-products.

6. *Provide separate reactor for recycle streams.* Recycling by-product and waste streams is an excellent technique for reducing waste, but
often the ideal reactor conditions for converting recycle streams back to
usable products are considerably different from conditions in the primary reactor. One solution is to provide a separate, smaller reactor for
handling recycle and waste streams. Then temperatures, pressures,
and concentrations can be optimized in both reactors to take maximum
advantage of the reaction kinetics. See Fig. 10.3.

7. *Consider a different reactor design.* The classic stirred-tank
back-mix reactor is not necessarily the best choice. A plug flow reactor
offers the advantage that it can be staged, and each stage can be run at
different conditions (especially temperature), closely controlling the

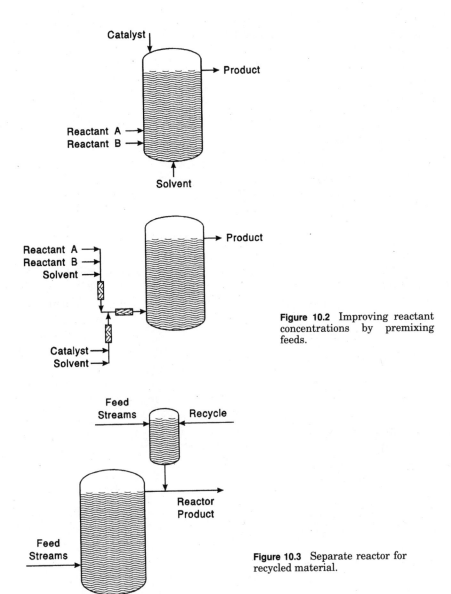

Figure 10.2 Improving reactant concentrations by premixing feeds.

Figure 10.3 Separate reactor for recycled material.

reaction for optimum product mix (and minimum waste). Many innovative hybrid designs are possible.

8. *Improve control.*   For a given reactor configuration, there is one set of operating conditions that is optimum at any given time. The control system should know those conditions and make them occur, with little fluctuation. Such control may be complex, particularly in the case of batch reactors, but it can yield major improvements. In less sophisti-

cated systems, simply stabilizing reactor operation frequently reduces the formation of waste products.

Advanced computer control systems are capable of responding to process upsets and product changes swiftly and smoothly, producing a minimum of unwanted by-products (see the section Process Control).

Heat Exchangers

Heat exchangers can be a source of waste, especially with products that are temperature-sensitive. There are a number of techniques for minimizing the formation of waste products in heat exchangers. All reduce the amount of fouling.

1. *Desuperheat plant steam.* High-pressure plant steam may contain several hundred degrees of superheat. Desuperheating the steam when it enters a process (or just upstream of an exchanger) reduces tube wall temperatures and actually increases the effective surface area of the exchanger because the heat-transfer coefficient of condensing steam is 10 times greater than that of superheated steam.

2. *Use lower-pressure steam.* When plant steam is available at fixed pressure levels, a quick expedient may be to switch to steam at a lower pressure, reducing tube wall temperatures.

3. *Install a thermocompressor.* Another method of reducing the tube wall temperature is to install a thermocompressor, shown in Fig. 10.4. These relatively inexpensive units work on an ejector principle, combining high- and low-pressure steam to produce an intermediate-

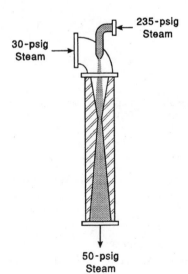

235-psig
Steam

30-psig
Steam

50-psig
Steam

Figure 10.4 Thermocompressor principle.

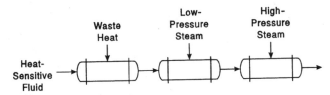

Figure 10.5 Staged heating.

pressure steam. Variable throat models are available that operate as control valves, automatically mixing the correct amounts of high- and low-pressure steam.

4. *Use on-line cleaning techniques.* On-line cleaning devices such as recirculated sponge balls or reversing brushes have been on the market for many years. In addition to reducing exchanger maintenance, they keep the tube surfaces clean so that a lower-temperature heat source can be used.

5. *Use scraped-wall exchangers.* Scraped-wall heat exchangers consist of a set of rotating blades inside a vertical, cylindrical jacketed column. They can be used to recover salable products from viscous streams. A typical application is to recover monomer from polymer tars.

6. *Use staged heating.* If a heat-sensitive fluid must be heated, staged heating can lessen degradation. Begin, e.g., with low-level waste heat, then use low-pressure steam and finally desuperheated intermediate-pressure steam. See Fig. 10.5.

7. *Monitor exchanger fouling.* Exchanger fouling does not necessarily occur steadily. Sometimes, an exchanger fouls because of some change in conditions or upset elsewhere in the process. The cause of fouling can be pinned down and reduced by continuously monitoring the fouling factor.

8. *Use noncorroding tubes.* Corroded tube surfaces foul more quickly than noncorroded tube surfaces. Changing to noncorroding tubes can significantly reduce fouling.

Pumps

Pumps do not usually contribute to waste except in two areas:

1. *Recover seal flushes and purges.* Each pump seal flush and purge needs to be examined as a possible source of waste. Most can be recycled to the process with little difficulty.

2. *Use sealless pumps.* Leaking pump seals lose product and create environmental problems. Using can-type sealless pumps or magnetically driven sealless pumps eliminates these losses.

Furnaces

Furnace technology is constantly evolving. Furnace manufacturers should be contacted for the latest techniques in optimizing furnace operation and reducing tar formation.

1. *Replace the coil.* In certain applications, significant improvements can be made by replacing the existing furnace coil with one having improved design features (e.g., tubes with low residence time or designed for split flow). Although it may not be practical to replace an undamaged coil, alternative designs should be considered whenever replacement becomes necessary.

2. *Replace furnace with intermediate exchanger.* Another option is to eliminate direct heating in the furnace (which necessarily exposes the heated fluid to high tube wall temperatures) by using an intermediate heat-transfer medium (e.g., Dowtherm®). This is illustrated in Fig. 10.6.

3. *Use existing steam superheat.* Even though the temperature required may be above the saturation temperature of plant steam, there may be sufficient superheat available in the steam to totally eliminate the need for a furnace. This is illustrated in Fig. 10.7.

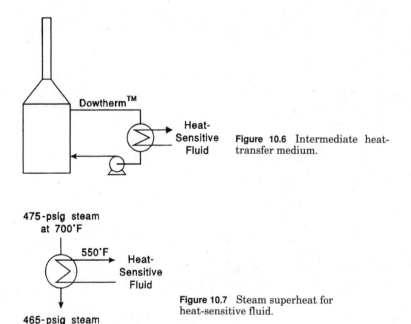

Figure 10.6 Intermediate heat-transfer medium.

Figure 10.7 Steam superheat for heat-sensitive fluid.

Distillation Columns

Distillation columns typically contribute to waste in three ways:

1. By allowing impurities to remain in a product. The impurities ultimately become waste. The solution is better separation. In some cases, it may be desirable to exceed the normal product specifications.

2. By forming waste within the column itself, usually because of high reboiler temperatures which cause polymerization. The solution is to find ways to lower column temperatures.

3. By inadequate condensing, which results in vented or flared product. The solution is to improve condensing.

In the following paragraphs, we will look at some column and process modifications that reduce waste by attacking one or more of these three problems.

1. *Increase the reflux ratio.* The most common way of improving separation is simply to increase the reflux ratio. This raises the pressure drop across the column and increases the reboiler temperature (using additional energy), but is probably the simplest solution if the column capacity is adequate.

2. *Add a section to column.* If a column is operating close to flood, a new section can be added to increase capacity and improve separation. The new section can use trays, regular packing, or high-efficiency packing. It need not be the same as the original portion of the column.

3. *Retray or repack column.* Another method of increasing separation is to retray or repack part or all of the column. High-efficiency packing is available and has the added advantage that it lowers pressure drop through the column, decreasing the reboiler temperature.

4. *Change feed tray.* Do not overlook changing the feed tray. Many columns are built with multiple feed trays, but the valving is seldom changed. In general, the closer the feed conditions are to the top of the column (high lights concentration and low temperature), the higher the feed tray; the closer feed conditions are to the bottom of the column (high heavies concentration and high temperature), the lower the feed tray.

5. *Insulate.* Good insulation is necessary to prevent heat losses. Poor insulation requires higher reboiler temperatures and allows column conditions to fluctuate with weather conditions.

6. *Improve the feed distribution.* The effectiveness of feed distributors (particularly in packed columns) needs to be analyzed to be sure that distribution anomalies are not lowering overall column efficiency.

7. *Preheat column feed.* Preheating the feed should improve column efficiency. Supplying heat in the feed requires lower tempera-

tures than supplying the same heat to the reboiler and reduces the reboiler load. It can often be done by cross-exchange with other process streams.

8. *Remove overheads product from tray near top of column.* If the overheads product contains a light impurity, it may be possible to obtain a higher-purity product from one of the trays close to the top of the column. A bleed stream from the overheads accumulator can be recycled back to the process to purge the column of lights. Another solution is to install a second column, which may be expensive but well worth doing.

9. *Increase the size of the vapor line.* In low-pressure or vacuum columns, pressure drop is especially critical; installing a larger vapor line reduces pressure drop and decreases the reboiler temperature.

10. *Modify reboiler design.* It may be desirable to use a falling film reboiler, a pumped recirculation reboiler, or high-flux tubes to minimize product degradation.

11. *Reduce reboiler temperature.* The same general temperature reduction techniques discussed earlier (in the section on heat exchangers) such as desuperheating steam, using lower-pressure steam, installing a thermocompressor, and using an intermediate heat transfer fluid, also apply to the reboiler of a distillation column.

12. *Lower column pressure.* Reducing the column pressure will also decrease the reboiler temperature and may favorably load the trays or packing as long as the column stays below flood. The overheads temperature, however, will also be reduced, which may create a condensing problem (see the following paragraph).

13. *Improve overheads condenser.* If overheads are lost because of an undersized condenser, consider retubing, replacing the condenser, or adding a supplementary vent condenser to minimize losses. It may also be possible to reroute the vent back to the process (if the process pressure is stable). If a refrigerated condenser is used, be sure to keep the tubes above 32°F if there is any moisture in the stream.

14. *Improve column control.* The comments in the Reactor section about closer process control and controlling at the right point apply to distillation columns as well as reactors.

15. *Forward vapor overheads to next column.* If the overheads stream is sent to another column for further separation, it may be possible to use a partial condenser and introduce a vapor stream to the downstream column.

Before any equipment modifications are undertaken, it is recommended that a computer simulation be done and that a variety of operating conditions be examined. If the column temperature or pressure changes, equipment ratings should also be reexamined.

Piping

Something as seemingly innocuous as plant piping can sometimes cause waste, and simple piping changes can result in a major reduction of waste. Consider the following piping changes to a process:

1. *Recover individual waste streams.* In many plants, various waste streams are combined and sent to a waste treatment facility, as shown in Fig. 10.8. Each waste stream needs to be considered individually. The nature of the impurities may make it possible to recycle or otherwise reuse a particular stream before it is mixed with other waste streams and becomes unrecoverable. Stripping, filtering, drying, or some other type of treatment may be necessary before the stream can be reused.

2. *Avoid overheating lines.* If a process stream contains temperature-sensitive materials, the quantity and temperature level of line and vessel tracing and jacketing need to be reviewed. If plant steam temperatures are too hot, a recirculated warm fluid can be used to prevent freezing. Electric tracing is also an option.

3. *Avoid sending hot material to storage.* If a temperature-sensitive material is sent to storage, it should be cooled first. If this is uneconomical because the stream from storage needs to be heated when it is used, simply piping the hot stream directly into the suction of the storage tank pump may solve the problem adequately. Make sure that the storage tank pump can handle hot material without cavitating. This is illustrated in Fig. 10.9.

4. *Eliminate leaks.* Leaks can be a major contributor to a plant's overall waste, especially if the products cannot be seen or smelled. A

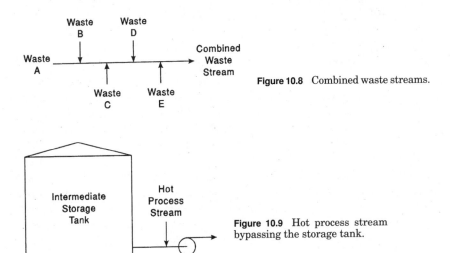

Figure 10.8   Combined waste streams.

Figure 10.9   Hot process stream bypassing the storage tank.

good way to document leaks is to measure the quantity of raw materials that must be purchased to replace "lost" streams (e.g., the amount of refrigerant purchased).

5. *Change metallurgy.* The type of metal used for vessels or piping may be causing a color problem or may be acting as a catalyst for the formation of unwanted by-products. If this is the case, change to more inert metals (see the next paragraph).

6. *Use lined pipes or vessels.* Using lined pipes or vessels is often a cheap alternative to using exotic metallurgy. A variety of coatings are available for different applications.

7. *Monitor vent and flare systems.* The measurements need not be highly accurate, but should give a reasonable estimate of how much product is lost and when losses occur. It is often worthwhile to install whatever piping is necessary to recover products that are vented or flared. Additional purification may be needed before the recovered streams can be reused.

Process Control

Modern technology allows us to install highly sophisticated computer control systems that respond far more quickly and accurately than human beings. We can use that capability to reduce waste. Some suggestions are summarized on the following pages.

1. *Improve on-line control.* Good process control reduces waste by minimizing cycling and improving a plant's ability to handle normal changes in flows, temperatures, pressures, and compositions. Statistical quality control techniques help plants analyze process variations and document improvements. Sometimes additional instrumentation or on-line stream monitors (e.g., gas chromatographs) are necessary, but good control optimizes process conditions and reduces plant trips, often a major source of waste.

2. *Optimize daily operation.* If a computer is incorporated into the control scheme, it can be programmed to analyze the process continuously and to optimize operating conditions. If the computer is not an integral part of the control scheme, off-line analyses can be performed and used as a guide for setting process conditions.

3. *Automate start-ups, shutdowns, and product changeovers.* Huge quantities of waste can be produced during plant start-ups, shutdowns, and product changeovers, even when such events are well planned. Programming a computer to control these situations brings the plant to stable operating conditions quickly and minimizes the time spent generating off-specification product. Further, since minimal time is spent in undesirable running modes, equipment fouling and damage are reduced.

4. *Program plant to handle unexpected upsets and trips.*   Even with the best control systems, upsets and trips occur. Not all can be anticipated, but operators who have lived with the plant for years probably remember most of the important ones and know the best ways to respond. With computer control, optimum responses can be preprogrammed. Then, when upsets and trips occur, the computer takes over, minimizing downtime, spills, equipment damage, product loss, and waste generation.

Miscellaneous Improvements

In addition to the ideas already discussed, a number of other miscellaneous improvements can be made to reduce waste:

1. *Avoid unexpected trips and shutdowns.*   A good preventive maintenance program and adequate sparing of equipment are two keys to minimizing trips and unplanned shutdowns. Another key is to provide an adequate warning system for critical equipment (e.g., vibration monitors). Plant operators can be extremely helpful by reporting unusual conditions so that minor maintenance problems get corrected before they become major and cause a plant trip.

2. *Use waste streams from other plants.*   Within a chemical complex, each plant's waste streams should be clearly identified. The quantity and quality of these streams should be documented, including the presence of trace quantities of metals, halides, or other impurities that render a stream useless as a raw material. This list should be reviewed by all plants in the complex to determine if any are suitable as feedstocks.

3. *Reduce the number and quantity of samples.*   Taking frequent and large samples can generate a surprising amount of waste. Many plants find that the quantity can be reduced or that the samples can be returned to the process after analysis.

4. *Recover product from tank cars and tank trucks.*   Product vented or drained from a tank car or tank truck (especially those dedicated to a single service) can often be recovered and reused.

5. *Use removable insulation.*   When conventional insulation is removed from equipment it is typically scrapped and sent to landfill. A number of companies manufacture reusable insulation. Their products are particularly effective on equipment where the insulation is removed regularly in order to perform maintenance (e.g., heat-exchanger heads, manways, valves, transmitters).

6. *Maintain external painted surfaces.*   Even in plants handling highly corrosive materials, external corrosion can be a major cause of pipe deterioration. Piping and vessels should be painted before being insulated, and all painted surfaces should be well maintained.

7. *Find a market for waste products.*   Converting a waste product to a salable product may require some additional processing, but this can be an effective means of reducing waste. Converted product, however, should not create a waste problem for the customer.

Conclusions

As a closing thought (and additional impetus to action), recognize that waste reduction is not synonymous with lower profits. You might be surprised to learn that nearly every idea listed in this chapter had a return on investment greater than 30 percent! Most paid for themselves in less than 1 year. This is discussed in more detail in Chap. 22.

Reducing waste is not just responsible management, it is good business!

Note: A version of this material was originally presented at "Pollution Prevention for the 1990's: A Chemical Engineering Challenge," December 1989, Washington, D.C., and was subsequently published in Hydrocarbons Processing (March 1990) and Chemtech (August 1990).

11

Unit Operations Database for Transferring Waste Minimization Solutions

Richard L. Smith, Jr., Ph.D.
Department of Chemical Engineering
Tohoku University
Sendai, Japan

Jamil A. Khan, Ph.D., P.E.
Department of Mechanical Engineering
University of South Carolina
Columbia, South Carolina

Introduction

The objective of this chapter is to introduce a practical method for developing alternative solutions to waste minimization problems. The method is based on a database of industrial solutions compiled from case studies, the EPA, and consultant reports of numerous industries. Applicable solutions for a given problem are located and related through the common unit operations between industrial categories. These solutions are compiled into a database program that is menu-driven. This, in turn, can help the user identify options which may merit more detailed study by some of the other techniques given in this text.

This methodology is embodied in a software system developed by the authors and known as *unit operations waste minimization,* or UO-WM. Further details on the system structure and the unit operations tree used in the program can be found in Refs. 1 and 2.

Basis and Scope

The procedure is limited to the steps described by Douglas[3] that lead to the definition of the input/output structure of a flowchart. The methodology is applicable to both the design phase of chemical process development and retrofit. However, to date, it has been applied almost exclusively during waste minimization assessments of established chemical and manufacturing processes in South Carolina, in which management was seeking methods to reduce the waste generated.

Unit Operations Approach

Serageldin et al.[4] used a *unit operation system* (UOS) to define industrial cleaning operations, such as parts cleaning, and to track volatile organic compound (VOC) emissions (see also Chap. 20). Defining a UOS provided a well-established boundary for which VOC emissions could be accounted for through material balances. In tracking VOC emissions, the reduction opportunities could be more readily realized.

The EPA[5] has developed a software tool to aid in process analysis for pollution prevention called Strategic WAste MInimization Initiative (SWAMI). The user inputs the unit operations and a process flow diagram is generated. A number of parameters are input by the user such as process waste streams and material flow rates. The system outputs a statement of relevant facts and a list of keywords to guide the user toward more detailed information for developing pollution prevention alternatives.

The methodology described in this chapter and embodied in the UO-WM system differs from the above approaches in that it provides detailed information on pollution prevention solutions currently used in industry, along with a ranking according to economic and pollution prevention factors. The design of the system is shown in Fig. 11.1.

In order to make solutions more accessible or generic to all types of industries, the unit operation was chosen as the basis to define waste generation as opposed to the process or industry that generated the waste. The first step in the procedure therefore involves defining the unit operation that generates the waste for a given solution.

Once the unit operations are defined, a unit operations tree can be constructed to categorize the operations. This is an extensive listing of all unit operations performed in industry. It is divided into gas-liquid, liquid-solid, and numerous other operations. These are further categorized

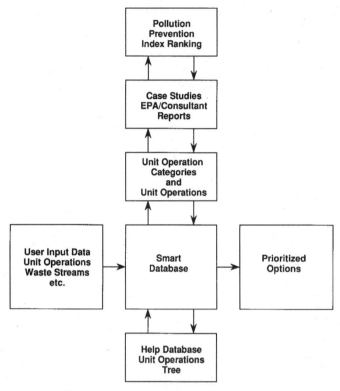

Figure 11.1 Database components.

into six broad categories where wastes are generated, namely, reactions, separations, mechanical, housekeeping, transportation, and laboratory.

Solutions Database

Information on pollution prevention solutions is available from case studies, EPA reports, consulting reports, and scientific literature. The solutions can be characterized by the waste-generating unit operation. Economic information, such as the capital cost and payback period, provides a basis for prioritizing solutions. A sample solution database record is shown in Table 11.1. A profile of the solutions database developed for UO-WM is given in Table 11.2 according to the information source and type. Solutions according to Standard Industrial Classification (SIC) codes are given in Table 11.3.

As shown in Table 11.1, each record of the solution database consists of 20 fields for technical, economic, archival, and other information. The solution is given in field 5. Fields 6 and 7 provide the source of the solution. The industry which developed the solution is given in field 19.

TABLE 11.1   Sample Solution Database Record

1. Unit operation:	Plating
2. Qualifier:	Metal equipment/parts
3. Waste stream:	Chromium/drag-out
4. Unit operation category:	Reaction
5. Solution:	Decrease the concentration of chromium plating solution from 3 to 29 oz/gal with no reduction in quality
6. Manual:	Proven profits from pollution prevention
7. Manual pages:	117–129
8. Source reduction:	Yes
9. Recycling:	No
10. Waste treatment:	No
11. Ease of implementation:	1
12. Percent reduction:	9%
13. Capital cost:	No cost
14. Payback period:	0
15. Depth of solution:	1000
16. Solution unit operation:	Plating
17. Solution of unit operation category:	Reaction
18. Solution unit operation qualifier:	Solid-liquid
19. Solution SIC code:	34XX
20. Pollution prevention index:	1005009591000

TABLE 11.2   Profile of Solutions Database

Number of solutions	277
Number of source reduction solutions	96
Number of recycling solutions	162
Number of waste treatment solutions	21
Number of solutions in the following categories:	
Reaction	112
Separations	38
Mechanical	98
Laboratory	2
Housekeeping	21
Transportation	6
Number of solutions from EPA reports	87
Number of solutions from company case studies and journal articles	133
Number of solutions from consulting reports	57
Number of solutions with complete economic information	149
Number of solutions with complete technical information	101
Number of solutions with complete economic and technical information	58

TABLE 11.3   SIC Codes of Solutions/Frequency Count

SIC code	Industry	Frequency
02XX	Agricultural production	1
10XX	Metal mining	6
20XX	Food and kindred products	4
22XX	Textile manufacturing	7
25XX	Furniture and fixtures manufacturing	5
26XX	Paper and allied products	1
27XX	Printing and publishing	6
28XX	Chemicals and allied products	65
30XX	Rubber and miscellaneous plastic products	1
31XX	Leather and leather products	2
32XX	Stone, glass, and concrete products	1
33XX	Primary metal industries	4
34XX	Fabricated metal products	89
35XX	Industrial and commercial machinery and computer equipment	5
36XX	Electronic and other electrical equipment	27
37XX	Transportation equipment	8
38XX	Measuring, analyzing, and controlling instruments	8
39XX	Miscellaneous manufacturing industries	2
45XX	Air transportation	2
49XX	Electric, gas, and sanitary services	13
51XX	Wholesale trade–nondurable goods	2
55XX	Automotive dealers and gasoline service stations	1
73XX	Business services	6
75XX	Automotive repair, services, and parking	2
76XX	Miscellaneous repair services	1
91XX	Executive, legislative, and general government	1
97XX	National security and international affairs	2

Beginning from a solution, first one determines the waste stream, the unit operation category, and the unit operation that generates the waste. In this example, the waste stream is chromium/drag-out, the unit operation category is *reaction,* and the unit operation is *plating.* To further describe the process, a qualifier is added. The solution is in a solid-liquid reaction category, indicated by fields 17 and 18. Since the solution does not involve alternate technology, field 16 is identical to the unit operation generating the waste.

Fields 9 through 15 are factors used to evaluate a solution. These factors are used to develop a pollution prevention index as described in a later section. The value of this index is used to rank solutions for a given flow sheet with the higher index value being the preferable solution. The value of the index is given in field 20.

The information in Table 11.2 shows that most solutions fall into the reaction, separation, or mechanical categories. A majority of the solutions come from industrial case studies. Table 11.3 gives the frequency by industry, and it can be seen that the primary solutions come from the chemical (28XX), plating (34XX), and electronic (36XX) industries. Of note is that the database has considerable breadth as solutions are drawn from 27 different industries.

Pollution Prevention Index

Many solutions exist for each unit operation. To evaluate a solution, many authors have used economic objective functions with specific technologies.[6,7] The disadvantage of such an approach is that treatment may be emphasized more than source reduction or recycling. For example, this could cause simple solutions such as "preinspect parts before plating" to be overlooked, when one is examining how to recover metals from a chemical strip operation. To overcome this difficulty and still allow location of specific technologies, a *pollution prevention index* (PPI) was developed for each solution. The details of this index are shown in Table 11.4.

The pollution prevention index is made up of eight weighted factors. The factors are determined from each waste minimization solution. Each factor is assigned a weight. The first three factors—source reduction, recycling, and waste treatment—are given the greatest weights according to the EPA pollution prevention hierarchy.[8] Any source reduc-

TABLE 11.4   Pollution Prevention Index

Source reduction (SR):	Yes = 1/no = 0
Recycling (R):	Yes = 1/no = 0
Waste treatment (WT):	Yes = 1/no = 0
Ease of implementation (EI):	1 = procedure change
	2 = retrofit equipment
	3 = new equipment
	4 = higher-purity solvent
	5 = material substitution
Percentage reduction (PR):	0–100%
Capital cost (CC) (1993$):	5 = no cost ($0)
	4 = low (<$15,000)
	3 = moderate ($15,000 < cost < $50,000)
	2 = High ($50,000 < cost < $150,000)
	1 = Very high (cost > $150,000)
Payback (PB):	In years 0 to 9
Depth of the solution (DS):	1000 = company case study
	0100 = EPA case study
	0010 = consultant report
	0001 = option

Pollution prevention index:

$$PPI = SR \times 10^{11} + R \times 10^{10} + WT \times 10^9 + (6 - EI) \times 10^8 + PR \times 10^5 + CC \times 10^4 + (9 - PB) \times 10^3 + DS$$

tion technique is preferred over a recycling or treatment technique. (See Chap. 4 for further discussion of the pollution prevention hierarchy.)

Five other factors are used to describe a waste minimization solution in terms of what is involved in implementing the solution (ease of implementation), the percentage of reduction that can be expected (percentage reduction), how much the solution will cost (capital cost), and on what scale the solution has been implemented (depth of solution). For ease of implementation, procedural changes are considered to be easier than material substitution. For the depth of solution, company case studies are preferred to consultant reports or options and recommendations made in the field or literature, because it means that the solutions are practical and economical and have been implemented to some degree by actual companies. No attempt was made to distinguish solutions proposed by large companies versus small companies, although capital cost is a measure of whether a solution is affordable for a smaller firm.

Each factor in Table 11.4 is combined with weights to produce a single value that can be used to rank the solutions. The user can change these weights to suit individual preference.

Analysis of Flow Sheets

Before we proceed further into the analyses, note that many of the alternatives that are located by unit operations may use different chemicals or may not be directly applicable to the process of interest to the user. However, in many cases, the solutions identified can provide a seed for an idea or alternative. For example, a process engineer manufacturing specialty chemicals might put the unit operation *mixing* into the database. This might pull up all the painting industry solutions, such as "use Teflon-lined vessels to reduce solvent cleaning wastes." Although not directly applicable to specialty chemicals, Teflon-lined mixing tanks could allow for a reduction in solvent waste generation provided the materials and chemicals were compatible. The common feature that allows transfer of solution between the two industries is the unit operation of mixing. Each solution needs to be viewed with the proper perspective.

Using Toxic Release Inventory Data

Toxic release inventory (TRI) data,[9] although not required for this procedure, offer useful adjunct information. They can provide the user with a list of other companies which have similar waste streams. Recently, the South Carolina regulatory agency has developed a very user-friendly PC-based TRI database. Through our waste minimization visits, we have found that companies having similar waste streams can have common unit operations. In many cases they are willing to share techniques or information (e.g., the type of activated carbon used in an adsorption solvent recovery unit, or other parameters) with other companies. Through communication, it is possible to share public-domain techniques which can lead to waste reduction. The power of this technique should not be underestimated.

To provide the user with a demonstration of the use of such a TRI database, let's take an example. Suppose a company has a problem with methanol air releases. A search "methanol" with the TRIS1991 system[9] turns up numerous *local* companies which release methanol into the air, land, water or transfer the material off site. The companies with air releases of methanol can be pulled, and technical factors such as recovery efficiencies can be examined. This information can serve as a basis for company meetings.

With the TRI data, companies which transfer the material can also be contacted concerning the end use of the methanol. Similarly, searches can be carried out for companies with similar products in which methanol releases are low or nonexistent. This indicates that the company either has found a superior chemical substitute (which may be proprietary) or has an effective recycling or recovery method.

Textile Screen Company Profile

Figure 11.2 gives a plant layout for a South Carolina textile screen company which generates 44,000 kg/yr of F006 electroplating sludge. The company specializes in the manufacture of nickel screens used for textile printing. In the manufacture, steel rolls are copper-plated, milled, knurled, and then impregnated with a nonconductive epoxy material. The copper roll is plated with chromium to form a base which has numerous tiny pinholes. Nickel is plated to the chromium roll. The nickel is more reactive than the chromium and does not adhere to the roll. The final result is a textile screen which can be coated by customers with light-sensitive material for printing designs onto fabrics.

As can be seen in Fig. 11.2, the company mixes all waste streams in a below-ground channel before sending the wastewater to waste treatment. As part of a waste minimization assessment, we made recommendations using our system and our experience. Figure 11.3 shows the analysis made with our system for a few representative operations.

The primary option recommended with our system is to increase the contact time between the workpiece and the rinse solution. This has the effect of reducing the water consumption and increasing rinse efficiency. Options ranked 2 and 7 also reduce rinse water requirements. Other options include the use of an atmospheric evaporator to evapo-

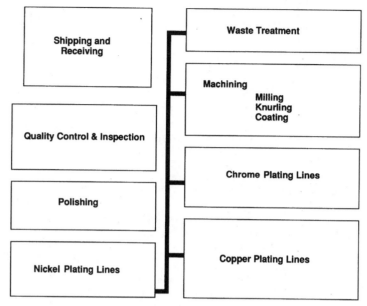

Figure 11.2 Plant layout for a textile screen company.

rate rinse water from the nickel electroplating bath or electrodialysis reversal to recover plating chemicals from the rinse water to reduce sludge generation. These options can lead to an 80 percent reduction in sludge generation. For waste treatment, use of sodium borohydride instead of lime/ferrous sulfate can reduce the plating sludge by 40 percent. Options are ranked according to the PPI of each solution.

Plating Company Profile

Figure 11.4 gives a plant layout of a South Carolina plating firm which generates 3200 kg/yr of F006 plating sludge. The company specializes in cadmium, nickel, and chromium plating and aluminum anodizing. Manual plating is performed on a contract basis. Since the pieces vary widely in shape and condition, a large number of pretreatments are required to remove carbonized welding fluxes, oils, and other contaminants. Invariably, readily soluble metals such as copper show up in the plating baths and cause quality problems. We will limit our discussion here to the chromium plating line.

The chrome-plating process can be broken down into a number of unit operations which are shown in Fig. 11.5. In the figure, parts to be plated move from top to bottom and are contacted in each of the tanks. Parts are first soaked and cleaned with a soap solution. The soap solution is removed with a single stagnant rinse. The parts travel to a bright dip-

UNIT OPERATIONS	SOLUTIONS	Overall Ranking According To Pollution Prevention Index (1 = BEST)
Machining	Use synthetic metal working fluids to increase the fluid life; tramp oils do not contaminate synthetic oils easily	2
	Recycle machining fluids by gravity pressure and vacuum filtration, recycle the machining fluids	7
Bright Dipping	Use sulfuric acid and hydrogen peroxide pickling process instead of a cyanide acid and chromic acid dip to reduce the amount of waste generated	3
Plating	Use an atmospheric evaporator to evaporate waste-water from the electroplating bath	6
	Use electrodialysis reversal to recover plating chemicals from rinse water to reduce sludge generation	8
	Use reverse osmosis to recover nickel plating solution from drag-out; this can reduce sludge generation by 80%	9
Rinsing	Increase the contact time between the workpiece and the rinse solution for more efficient rinsing	1
	Recycle the rinse water; use as a rinse after bright dipping or alkaline cleaner	4
	Use spray rinsing above the process tank to reduce drag-out losses	5
Waste Treatment	Use sodium borohydride instead of lime/ferrous sulfate to reduce RCRA F006 waste by 40%	10

Figure 11.3 Analysis of selected plating operations of a textile screen company.

ping operation to remove gross scale. For this operation, sulfuric acid–hydrogen peroxide solutions were used. The bright dipping operation is followed by a single stagnant rinse. The parts are then chrome-plated and rinsed in a three-stage countercurrent rinse sequence.

The unit operations for this example are rinsing, bright dipping, and chrome plating. The possible qualifiers are metal parts, equipment, and tanks. The waste streams are chromium-water solutions, copper-water solutions, and sulfuric acid–hydrogen peroxide. This information can be used to locate relevant waste minimization solutions.

For each unit operation, the program locates numerous solutions which are ranked automatically by the pollution prevention index.

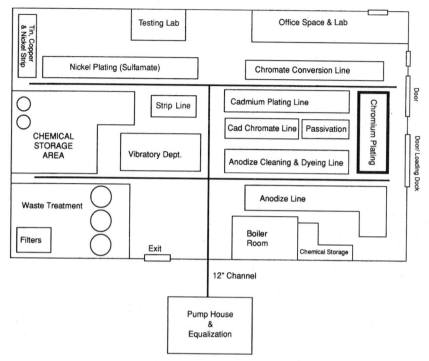

Figure 11.4   Plant layout for a plating company (job shop).

According to the priorities set forth in Table 11.4, source reduction techniques are ranked higher than recycling techniques, and among those, solutions that are easiest to implement and give the highest percentage of waste reduction are ranked accordingly.

For example, the program finds that copper can be recycled with ion exchange in the bright dipping operation. It also finds that the chromium drag-out can be reduced by decreasing the plating bath concentration from 32 to 29 oz/gal (240 to 217 kg/m^3). Each of these methods is a valid alternative for reducing hazardous waste. However, the second alternative should have higher priority because it is a source reduction technique that is easily implemented, whereas the ion-exchange technique is a recycling method that requires additional equipment.

Assuming that the company plates only chromium, implementation of solutions of rank 4 or higher would lead to an estimated minimum reduction of 88 percent in F006 plating sludge. This is estimated from the 9 percent for decreasing chromium bath concentration, 50 percent by adding surfactants to the plating bath, and 72 percent by modifying the rinse tank configuration. Each of these alternatives is a source reduction technique. For some techniques such as preinspecting parts

UNIT OPERATIONS	SOLUTIONS	Overall Ranking According To Pollution Prevention Index (1 = BEST)
Movement of Parts		
Soak and Clean		
Rinsing	Use EPA optimum rinse tank schemes	5
	Increase contact time between workpiece and solution	5
Bright Dipping	Continuously recirculate sulfuric acid/hydrogen peroxide dip through ion exchange to remove copper	10
	Remove saturated solution, precipitate copper sulfate and separate with cyclone separator	9
Rinsing	Use EPA optimum rinse tank schemes	5
	Use deionized water in stagnant rinsing to reduce drag-in	5
Chrome Plating	Preinspect parts to avoid plating defective parts	2
	Withdraw pieces slowly to reduce drag-out	3
	Decrease concentration of chromium plating bath from 32 oz /gal to 29 oz /gal	1
	Reduce anode-to-cathode distances	2
	Use sodium bisulfite in stagnant rinse following plating to reduce Cr(VI) to Cr(III)	12
	Use surfactants in the plating bath to reduce drag-out	7
Rinse	Recover chromium from the rinse water effluent with ion exchange	11
	Use EPA optimum rinse tank scheme:	
Rinse	a. Two drag-out rinse tanks, one overflow rinse tank	4
	b. Three drag-out rinse tanks, one overflow rinse tank	5
	Flow rate controls–rinse water on demand	6
Rinse	Use drainage boards between process and rinse tanks and route drippage back to process tank	8

Figure 11.5   Analysis of selected plating operations of a plating company.

or withdrawing pieces slowly, it is difficult to predict the percentage of reduction in hazardous waste. However, as indicated by our method, these approaches should be considered before technologies such as ion exchange or cyclone separation.

Chemical Company Profile

Figure 11.6 gives a plant layout of a South Carolina specialty chemical producer. The company generates 4600 kg/yr of D002 corrosive solid,

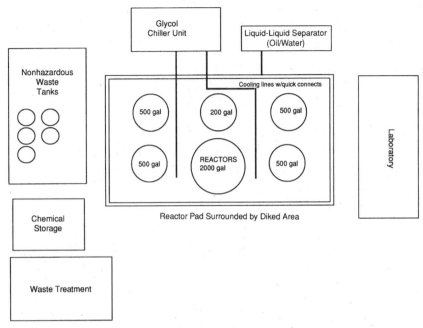

Figure 11.6   Plant layout for a specialty chemical company.

23,000 kg/yr of 6666 waste flammable liquid, and 4600 kg/yr of chromium-contaminated adsorbent and numerous other spent chemicals in small quantities.

Specialty chemicals are made by following known recipes from organic chemistry. These are scaled up according to chemical engineering design fundamentals, including mass and heat transfer as well as reactor configuration and agitation. After a reaction has reached the desired degree of completion, the product is removed from unreacted components by separation steps such as distillation, liquid-liquid extraction, crystallization, or adsorption.

A typical set of operations for oxidizing a secondary alcohol group with chromic acid can be broken down into the unit operations shown in Fig. 11.7. These include liquid-phase reaction, mixing, heating, refluxing, cleaning, other types of reactions, washing a solid with solvent, recrystallization, distillation (which is used to recycle some of the spent solvents), and waste treatment of the spent solvents.

A number of options were located with our system and are shown in Fig. 11.7 along with the solution ranking according to the PPI. The top-ranked solutions are source reduction options.

For example, in liquid-phase reaction, dedication of a product line that uses similar type solvents (e.g., chlorinated or nonchlorinated) can greatly reduce or eliminate cleaning requirements. In the design

UNIT OPERATIONS	SOLUTIONS	Overall Ranking According To Pollution Prevention Index (1 = BEST)
Material Transport	Use rinseable or recyclable drums with plastic liners instead of paper bags	14
Liquid Phase Reaction	Dedicate each reacting system to a specific product to avoid unnecessary cleaning	1
	Segregate spent chlorinated and nonchlorinated solvents and distill off-site for reuse	12
	Segregate hazardous and nonhazardous waste streams to reduce disposal costs	17
Mixing	Teflon-lined tanks to improve drainage and reduce cleaning	7
	Use chemicals in paste form instead of dry form	15
Heating		
Refluxing		
Cleaning	Alternate caustic and alkaline cleaners to reduce waste	2
	Adjust production schedule to minimize reactor cleaning	3
	Pump a fixed amount of cleaner into tank to reduce overall volume required	4
	Immediately clean mixing tank to reduce the need for caustic cleaners	4
	Use a high-pressure spray to reduce caustic usage	5
	Use volume-limiting hose nozzles	6
	Use steam cleaning of mixing tanks instead of batch boil	8
	Reuse waste solvent instead of using virgin solvents	9
	Use off-site distillation to recover MEK from spent solvents and reuse in cleaning operations	10
	Use noncontact cooling water for the cleaning operations	11
Washing		
Crystallization		
Waste Treatment	Skim oil and process chemicals from wastewater before treatment to reduce sludge	13
	Use sodium hydroxide to precipitate chromic acid solutions instead of adsorption onto Fuller's earth	18
	Use steam stripping to remove aliphatic halogenated solvents from wastewater, and recycle	16

Figure 11.7   Analysis of selected plating operations of a specialty chemical company.

phase, this consideration could allow justification of a second vessel based on waste generation alone!

In reactor cleaning, alternating caustic and alkaline cleaners reduces corrosive wastewater discharge by 50 percent. Similarly, using off-site distillation to recover spent solvents for use in cleaning operations can cut waste solvent disposal by as much as 60 percent. By following the recommendations of our system, it is possible to reduce waste flammable liquid by as much as 13,600 kg/yr and chromium waste by as much as 3900 kg/yr. Reductions in wastewater can also be significant.

Conclusions

This chapter describes an approach to pollution prevention founded on a database of pollution prevention solutions categorized by unit operation. The procedure is embodied in the UO-WM database system. It is a simple and effective method for analyzing process flow sheets for waste minimization. The system is practical because it is based on numerous industrial case studies and reports.

The pollution prevention index contains most of the important factors required to rank waste minimization solutions. Solutions can be ranked even when technical and economic information is incomplete. The PPI provides the user with the most important solutions based on source reduction.

Unit operations are an effective way to search and apply solutions between industrial categories. This approach can generate a very large list of possible solutions, not all of which are appropriate. However, the solutions can be efficiently screened by using additional qualifiers such as waste stream or part type.

The development of the solutions database is an ongoing process, and the procedure should become even more powerful as the database expands. It is planned to update the UO-WM database to accommodate the growing body of information.

Software

A shareware package containing the UO-WM program is available from the authors of this chapter. This will be updated periodically as new data become available.

Dr. Jamil Khan may be contacted at the Department of Mechanical Engineering, University of South Carolina, Columbia, South Carolina 29208. Dr. Richard Smith's current address is Tohoku University, Department of Chemical Engineering, Sendai, Japan T 980; email smith@scf.saito.che.tohoku.ac.jp.

To use UO-WM most effectively, the following are recommended:

1. UO-WM program with DBase III + or higher and an IBM PC compatible with 2 Mbytes of hard disk space

2. Process flow sheet with list of unit operations

3. TRIS (Toxic Release Inventory System) data, preferably in an electronic database format[9]

Items 1 and 3 are included in the shareware diskette. Item 1 is the UO-WM program. Item 2 is supplied by the user or from examples given in this chapter.

Acknowledgments

The authors wish to gratefully acknowledge financial support for this work from the Hazardous Waste Management Research Fund (HWMRF), administered by the South Carolina Universities Research and Education Foundation.

References

1. S. A. Slater, "A Unit Operations Approach to Waste Minimization: Development of a Rule Based System for Process Flowsheet Analysis," M.S. thesis, University of South Carolina, July 1993.
2. S. A. Slater, J. Khan, and R. L. Smith, Jr. "Transferring Waste Minimization Solutions between Industrial Categories with a Unit Operations Approach, I. Chemical and Plating Industries," *J. Environ. Sci. Health,* vol. 30, no. 2, 1995.
3. J. M. Douglas, "Process Synthesis for Waste Minimization," *Ind. Eng. Chem. Res.,* 31:238–243, 1992.
4. M. A. Serageldin, J. C. Berry, and D. I. Salman, "A Novel Approach for Gathering Data on Solvent Cleaning," presented at the 1992 EPA/A&WMA International Symposium on Measurements of Toxic and Related Air Pollutants," Durham, NC, May 3–8, 1992.
5. Environmental Protection Agency, *User's Guide: Strategic WAste MInimization Initiative (SWAMI) Version 2.0,* publication 625/11-91/004, Office of Research and Development, Washington, January 1992.
6. A. R. Ciric, "Process Synthesis for Waste Minimization," paper presented at the American Institute of Chemical Engineers annual meeting, Los Angeles, November 17–22, 1991.
7. R. F. Dunn and M. M. El-Halwagi, "Optimal Recycle/Reuse Policies for Minimizing the Wastes of Pulp and Paper Plants," paper presented at the American Chemical Society Symposium on Emerging Technologies for Hazardous Waste Management, Atlanta, October 1–3, 1991.
8. Environmental Protection Agency, *Waste Minimization Opportunities Assessment Manual,* publication 625/7-88/003, Cincinnati, 1988.
9. Copies of the latest TRI database system for South Carolina are available by contacting Michael Juras, South Carolina Department of Health and Environmental Control, 2600 Bull Street, Columbia, SC 29201, Attn: EPCRA Reporting.

Chapter

12

Hierarchical
Process Review for
Waste Minimization*

Alan P. Rossiter, Ph.D.
Linnhoff March, Inc.
Houston, Texas

Howard Klee, Jr., Sc.D.
Director of Environmental Affairs and Safety
Amoco Corporation
Chicago, Illinois

Introduction

Hierarchical process review provides a systematic approach to the problem of identifying appropriate process modifications to minimize waste generation. The method uses a logical framework for identifying process improvement options and evaluating heat and mass integration opportunities. This offers a structured alternative to the traditional ad hoc approach generally used in the conceptual phase of engineering design.

The technique brings together the science of the industrial process and the knowledge and creativity of engineers. It does not replace the need for creativity; rather, it serves to direct the creative process, and it allows improved emissions reduction opportunities to be identified

---

*Adapted with permission from "Apply Process Integration to Waste Minimization," *Chemical Engineering Progress,* January 1993. Copyright 1993, American Institute of Chemical Engineers. All rights reserved.

more easily and more rapidly than would otherwise be possible. Moreover, the procedure provides an overall process perspective, rather than focusing on localized effects. This enables benefits and costs to be assessed in terms of their impact on the process as a whole.

This chapter describes the hierarchical review approach and highlights new insights that have been incorporated within the overall methodology. Use of the procedure is illustrated by a study that was carried out at Amoco Oil Company's Yorktown, Virginia, refinery. In addition to reducing emissions, the process improvement options identified in the study offer other benefits, such as energy savings and improved raw-materials efficiency. All the schemes are based on existing technologies and equipment types, although in some cases the applications are novel.

Hierarchical Review

Hierarchical review has its origins in the hierarchical design concept.[1] The basic premise is that process designs evolve via a series of decisions, which provide successive levels of detail. By identifying the key decisions and implementing them in the correct order, it is possible to identify good design options with a minimum of effort and rework. Alternative designs for evaluation may also be generated by the procedure. At each level, an economic evaluation of the process can be carried out, providing a screening tool to eliminate nonviable processes without the need to expend excessive design effort.

The method has been adapted for retrofit applications.[2,3] In this case, rather than providing a logical sequence for making design decisions, the procedure provides a systematic framework for reviewing and improving existing installations. The approach has also been adapted for waste minimization, in both new design and retrofit applications. The objective in this case is the generation of new design or retrofit options for source reduction or recycling.[4] Where appropriate, end-of-pipe options are also identified. This form of the procedure is discussed below.

Outline of Methodology

The following hierarchy of decisions can be used for the analysis of new and existing plant designs with the objective of reducing emissions:

Level 1:   Processing mode: batch versus continuous
Level 2:   Input/output structure of the flow sheet
Level 3:   Recycle structure and product formation considerations
Level 4:   Separation system
Level 5:   Product drying

Level 6:   Energy systems

Level 7:   Equipment and pipework specifications

Options for the design of new processes and possible improvements for existing ones are considered sequentially in the order of the decision hierarchy, by using a combination of heuristics, questions, and numerical calculations. Rough economics for each option are evaluated as the option is identified, to avoid wasting effort on options that are clearly not viable.

Heat and material balances are needed as a basis for these evaluations. It is generally difficult to obtain accurate information on the amounts of materials released by processes, and conventional flow sheets often show these streams simply as "trace" flows. In practice, the hierarchical approach can be applied even where only very approximate data are available.

Use of the procedure in new plant design generally results in one or more "good" designs for the process; in retrofits, it generates a list of potential process modifications to reduce emissions economically.

Application of Methodology

At each level, certain specific issues need to be addressed. These are summarized below.

Level 1: Processing mode

Criteria influencing the fundamental choice between continuous and batch operation are considered, and implications for waste generation and disposal are evaluated. Two key differences in waste generation between batch and continuous processes are as follows:

- Waste streams from batch processes are generally intermittent and often include effluents created by cleaning between batches. Waste streams from continuous processes are generally relatively steady flows.

- Compositions and flow rates of waste streams leaving a batch process typically vary, whereas those of waste streams from continuous processes are generally fairly constant.

The greater variability of the wastes from batch processes tends to create more difficult waste management problems. For example, if a given total volume of wastes has to be handled, the instantaneous maximum flow rate is higher in a batch plant. Consequently, larger equipment items are required to handle the wastes in this type of process. Moreover, waste generation rates are often high during start-up and shutdown periods, and these occur most frequently in batch units. It follows that waste reduction factors generally favor continuous rather than batch processing.

A number of guidelines have been proposed for deciding between batch and continuous processes.[5] The factors and heuristics considered include

- *Production rate:* Under 500 t/yr, batch processing is almost invariable. Between 500 and 5000 t/yr, batch processing is common. At higher rates, continuous processing is usually preferred.

- *Product life:* Batch plants are better suited to products with short life spans, where a rapid response to the market is required.

- *Multiproduct capabilities:* If the unit is required to produce several similar products by using the same equipment, batch processing is usually preferred.

- *Process reasons:* A number of process-related factors may lead to batch processing's being preferred, e.g., cleaning requirements that necessitate frequent shutdowns, difficulty in scaling up laboratory data, or complicated process recipes.

The environmental considerations discussed above suggest that an additional heuristic should be added to this list:

- If potentially serious environmental problems are anticipated with a particular process, this favors the selection of a continuous unit.

In practice, some of the other factors mentioned above may still dictate that a batch operation is preferred. In this case, consideration should be given to "smoothing" intermittent or variable flow streams, e.g., by adding buffer storage capacity, to simplify processing and recovery of "waste" materials.

Level 2: Input/output structure

The input/output structure of the process is considered. Figure 12.1 shows one possible form this structure may have. Representing the process in a simplified input/output format provides a useful overview of the overall materials transformations taking place and helps to identify the individual waste streams that are generated.

The waste can often be related back to specific inputs to the process. For example, impurities in a feed stream may have to be removed in

Figure 12.1   Typical input/output structure.

the form of a process effluent. Using a purer feed material may eliminate the need for the effluent stream. A rather different example occurs in many separation operations, where an additional feed is provided for the sole purpose of assisting in the separation. Perhaps the most common form of this is live steam injection, where the resulting steam condensate generally becomes an impure waste stream that has to be eliminated from the process. This waste can sometimes be reduced or even eliminated by using a reboiler rather than live steam. A third example is the situation where one of the feeds is a diluent. It may be possible to recycle this within the process rather than supply it as a feed (see the discussion of recycle structures under level 3).

Key questions for level 2 include these:

- Can the feed quality be improved at low cost?

- Should impurities be removed before processing or be allowed to pass through the process?

- Can any input streams be eliminated?

- Can any "waste" output stream be used beneficially within the process plant or sold to an external customer?

Level 3: Recycle structure and product formation considerations

Here the process is broken down into its major component sections (typically reaction and separation), together with interlinking streams and recycles. One possible structure, typical of many petrochemical processes, is shown in Fig. 12.2. Where data are available, it is possible to assess the impact of reactor conditions on waste formation and to evaluate some of the major tradeoffs in the process. These typically include reactor conversion, separation/recycle cost, and waste generation due to by-product reactions.

Key questions for level 3 include these:

- Do any waste output streams contain feed or product material that could be recovered and recycled?

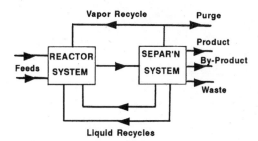

Figure 12.2 Typical recycle structure (vapor-liquid).

- Can reaction conditions be altered to minimize formation of waste by-products?
- Can waste by-products be recycled to extinction?

Level 4: Separation systems

Level 4 provides a more detailed analysis of separation options. A typical separation system structure for a vapor-liquid system is shown in Fig. 12.3. At this level, the impact of the separation technology (distillation, filtration, etc.) on waste generation is considered. Possible means of reducing emissions by using alternative separation techniques or by rearranging existing ones are evaluated. Interactions of the separation system with the reactor and recycle systems are also considered.

At level 4, the following questions should be addressed:

- Are any waste streams the result of poor or inappropriate separations?
- Can any wastes (especially hazardous wastes) be removed from process effluents by adding new separations? Typically this will

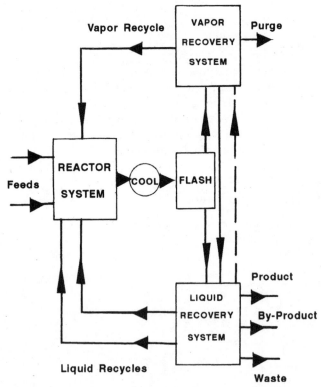

Figure 12.3 Typical separation system structure (vapor-liquid).

involve additional separation steps to recover the undesirable material from effluent, preferably for recycle or, if this is not possible, for safe disposal.

- Are there alternative separation technologies that could replace or supplement existing separations and reduce releases? Most industries have preferred separation technologies; e.g., distillation is the universal workhorse of the refining and petrochemical industries. Sometimes it may be possible to improve the quality of process effluents by using nontraditional separation equipment; e.g., membrane systems are showing promise in a variety of applications.

Level 5: Product drying

Level 5 is specific to solid-liquid processes and concerns product drying. However, some of the same issues are also relevant to other operations, such as fluid catalytic cracking, that use solid catalysts (see "Refinery Study" later in this chapter). Possible contributions to waste generation due to materials degradation in the dryer are assessed together with the control of materials losses from the dryer itself.

Often the mode of operation of the dryer is a direct cause of material losses; e.g., high air and particle velocities in flash dryers can lead to attrition of solid products, and thus to the creation of fines, which are difficult to remove from the airstream leaving the dryer. Selection of the most appropriate drying equipment is therefore important,[6] as it makes the cleanup of the air stream much easier. Air cleaning might require one or more of the following: cyclones, bag filters, wet scrubbers, electrostatic precipitators, or other proprietary equipment types.

Interactions of the dryer with other parts of the flow sheet may also be important in reducing net emissions. For example, if the materials recovered in air cleaning equipment are suitable for recycling to some part of the process, there is no need to dispose of them outside the process boundary. Recycling also improves raw-materials efficiency.

Key level 5 questions include these:

- Are there available dryer technologies that cause less product degradation (e.g., by gentler mechanical and/or thermal handling)?
- Can waste materials be removed from dryer off-gases and thus prevented from leaving the process and entering the environment?
- Can dryer wastes be recycled?

Level 6: Energy systems

At level 6, the energy systems required by the process are evaluated. The major tool for use at this level is *pinch analysis*. This provides powerful insights for process integration and is best known for its use in heat-exchanger network design. However, it has other far-reaching applications in reducing the environmental impact of industrial processes (see Chaps. 6 and 7). Benefits include these:

- Integration of end-of-pipe treatment steps into the rest of the process, to reduce capital and operating costs.

- Reduction of fuel firing and steam consumption. This, in turn, reduces the production of greenhouse gases and use of water treatment chemicals.

- Selection of appropriate utility systems to maximize cogeneration.

- Minimizing the required temperature levels for utility heating. Lower-temperature utilities (e.g., low-pressure steam) are generally cheaper and generate less net pollution than higher-temperature utilities, because

  - They are often produced in systems (e.g., backpressure steam turbines) in which mechanical work and/or electrical power is generated together with the heating medium. The total production of greenhouse gases in such cases is generally less than that of independent heating and power generation systems.

  - Low-pressure steam can sometimes be generated from waste heat within a production site. Exploiting opportunities to do this can significantly reduce fuel-firing requirements.

  - Lowering the temperature of heating media often reduces fouling in heat exchangers. This reduces exchanger cleaning requirements and the resultant waste disposal problems.

- Providing insights into possible modifications to reaction and separation processes for:

  - Reducing energy consumption
  - Reducing capital cost
  - Reducing emissions

These insights can then be used to explore trade-offs.

Questions for level 6 include these:

- How far can the energy consumption of the process be reduced economically?

- Can the temperature levels at which heat is delivered to the process be lowered?

- What fuels are used to provide heat for the process? Are cleaner fuels available and, if so, at what cost?

Level 7: Equipment and pipework

Level 7 provides an opportunity to evaluate and minimize fugitive emissions due to individual equipment items, valves, and pipework. The number of potential sources of fugitive emissions is minimized by "designing out" problematic equipment as much as possible. Emission characteristics are considered in the selection of critical equipment items.

Typical level 7 questions are as follows:

- Can the total number of equipment items and connections be reduced?

- Where there are alternative types of equipment for a particular service, which one offers the most favorable emission characteristics? Is the incremental cost justified by the reduced emissions?

- Can welded pipe be used in place of flanged pipe?

Heuristics

The preceding section contains a number of heuristics for design to minimize pollution and waste. As these areas mature and the growing body of literature is more fully digested, more heuristics will no doubt become available. The list of heuristics which follows includes a number that are mentioned in the preceding section, together with several based on other published data.[7,8]

1. Wherever possible, eliminate waste materials at their source. This can sometimes be achieved by switching raw materials.

2. Recycle waste materials within the process. Where this is not possible, explore the possibility of using these materials within other processes—preferably (but not necessarily) within the same operating facility.

3. If waste by-products are formed reversibly within a reaction process, they should be recycled to extinction. Where this is possible, it can totally eliminate a class of waste materials.

4. For all heating duties that require utilities, use the utility with the lowest practical temperature. This offers several potential benefits, including reduced heat-exchanger fouling and reduced utility costs.

5. Minimize the total number of main equipment items in the process, especially in areas that handle toxic materials. Also, minimize the number of pipework connections to and from equipment items.

6. Rapid, low-cost reductions in waste generation can often be achieved through changing set points or tightening control variations of key variables. Modifications to single equipment items (e.g.,

changing column internals) can also yield significant improvements with little capital expenditure.

7. Pollution prevention and control are generally more difficult and costly in batch operations than in continuous ones. Therefore, where practical, continuous processes are preferred.

A final observation on the use of heuristics: Haseltine[8] notes that the inherent flexibility of batch plants often makes raw material and product substitutions simpler in these processes. In this respect, pollution prevention is easier in batch processes—which is a direct contradiction of the final heuristic listed above. This typifies an underlying problem with heuristics: They are simply guidelines based on a body of experience, and they cannot be guaranteed to give the "right" answers in every case. They are, nevertheless, a very helpful way of getting rapid insights and generating options for evaluation.

Refinery Study

The hierarchical review procedure was applied to the crude unit and the fluid catalytic cracking (FCC) unit, together with the sour water system, at Amoco's Yorktown, Virginia, refinery. This work was carried out as part of the joint Amoco/EPA Pollution Prevention Project.[9] (See also Chap. 21.)

Most of the direct discharges from these units are liquids. However, vaporization of volatile organics from the sewers and water treatment facilities generates between 2 and 5 percent of the airborne hydrocarbon emissions in the refinery. Thus liquid releases may also have an impact on air quality. In addition, process fugitive emissions from valves, flanges, pump seals, etc., contribute 10 percent of the airborne hydrocarbon emissions; and vapors from storage tanks may contribute another 8 percent, depending on the type of seal in place.

The study resulted in a large number of ideas for reducing releases, some of which are summarized in Table 12.1. Most of these have been defined in much greater detail than can be shown here, and several were incorporated in the Amoco/EPA short list of pollution prevention options for the facility. All the potential projects are based on known technologies and available equipment types, although in several instances the applications are novel. The procedure enabled these ideas to be generated very quickly, with only limited process data. The use of the procedure in the FCC unit is described below.

The FCC unit produces fuel oil and components for gasoline blending by breaking long-chain "bottom of the barrel" hydrocarbons. The effluent from the FCC reactor contains a wide range of hydrocarbons, which are separated into the required products by distillation. An important

TABLE 12.1   Summary of Main Process Improvement Ideas

Description	Objectives
1. Replace steam ejectors with vacuum pumps	Reduce sour water flow and steam use
2. Two-stage desalting	Reduce brine flow and oil losses
3. Recover and recycle oil in desalter brine	Reduce hydrocarbon losses
4. Cool desalter brine	Reduce hydrocarbon flashing
5. Replace live steam with reboilers	Reduce sour water generation
6. Use stripped sour water for line washing	Reduce water import
7. Generate stripping steam from stripped sour water	Reduce sour water surplus, water import
8. Replace pump packings	Eliminate seal water losses
9. Recycle polygasoline water	Reduce feed to sour water system
10. Heat-integrate	Crude unit: Save up to 10% fuel. FCC: generate up to 30% more steam. Reduce NO_x, SO_x, and CO_x emissions
11. Minimize losses to flare	Improve raw-material efficiency; eliminate flare emissions
12. Recover flare gas as fuel	Improve raw-material efficiency; eliminate flare emissions
13. New solid recovery system for FCC regeneration	Reduce fines losses from regenerator
14. Recirculate stored DCO	Reduce sedimentation

feature of the process is the deposition of coke on the cracking catalyst. This deactivates the catalyst, which therefore requires continuous regeneration. Regeneration is a strongly exothermic process, and the heat is used to generate steam.

The FCC unit discharges considerable quantities of aqueous effluent. In addition, catalyst fines are lost from the regenerator as airborne dust. There are three major forms of water ingress (in addition to the small amount of water in the feed material):

- Stripping or fluidizing steam to the reactor and regenerator
- Wash water to vapor lines
- Steam stripping in light cycle oil (LCO) and decanter oil (DCO) strippers

Virtually all the water is removed as condensate or wash water effluent within the distillation system. This goes to sour water stripping. A very small amount of water is fed forward in product streams.

The hierarchical analysis of the FCC unit went as follows:

Level 1 (processing mode)

The process is continuous. There is no incentive to convert to batch operation.

Level 2 (input/output structure)

These issues were addressed:

1. Can stripping steam be eliminated in columns by using alternative separation technologies? (Refer to level 4.)

2. Can the quality of the feed be improved to eliminate or reduce the need for the vapor line washing system? This could not readily be answered, but alternative options were identified at level 3.

3. Can the steam consumption in the reactor be reduced so that less condensate will have to be removed from the distillation system? This would necessitate changing the reactor design and would probably not be a viable retrofit option.

4. Within the regeneration system, the loss of fines appears to be a function of air input rate (level 2) and cyclone design (level 4—separation system). At level 2, a reduction in air mass flow (e.g., by using oxygen enrichment) is identified as a possible means of reducing the discharge of fines. However, preliminary evaluation showed this to be less attractive than the other options.

Level 3 (reaction/recycle structure)

Two specific ideas came from the review of the reaction/recycle structure:

1. The reactor uses 26,000 lb/h (3.28 kg/s) of steam. This is provided from the utility steam system. If this could be replaced with steam generated from process water, the liquid effluent from the FCC unit could be greatly reduced. This would allow the condensate recovered within the distillation section (or other contaminated condensate) to be revaporized and recycled to the reactor section.

The steam generated in this way would contain volatile hydrocarbons from the recovered condensate, but this should not be a problem since it is returned directly to the process. A large blowdown rate would probably be needed, but the total volume of liquid effluent would be much less than in the existing plant, and a large part of the hydrocarbon content would be recovered.

Regeneration consumes more than 11,000 lb/h (1.39 kg/s) of steam. It may be possible to satisfy this duty also with "dirty steam," since the hydrocarbon content would be incinerated with the coke in the regenerator.

2. Used wash water is collected at several points and then purged from the process. If it could be recovered and recycled, or if recycled water from other sources could be used for washing in place of fresh water, the freshwater usage and wastewater generation could both be reduced by about 10,500 lb/h (1.32 kg/s).

Level 4 (separation system)

Three options were identified at this level:

1. Replace live steam stripping with reboilers. In practice, this would reduce water ingress by only 531 lb/h (0.07 kg/s), or less than 2 percent of the total, and so is unlikely to be worth doing. [However, the same idea applied to the sour water strippers would reduce water ingress by 23,000 lb/h (2.90 kg/s).]

2. Place additional oil-water separators downstream of existing condensate collection points. Separators could be, e.g., coalescers or strippers. Recover and recycle hydrocarbons.

3. Improve gas-solid separation downstream of the regenerator. This might simply require better cyclone and/or ductwork design, or electrostatic precipitation.

Level 5 (product drying)

There is no product drying in the FCC unit. However, recovery of catalyst fines downstream of the regenerator (see level 4, option 3) presents problems similar to those encountered in solids drying.

One possible means for reducing fines losses might be to develop a regenerator with less abrasive particle handling, which would produce fewer fines. However, this would not be a realistic or economical retrofit option. Another option investigated was the use of more attrition-resistant catalysts, but it turned out that the catalyst already in use was the best one available in this respect.

Level 6 (energy systems)

In steady operation, the FCC unit generates sufficient heat to be self-sustaining and requires no external heating utility. Much of the heat rejected by the unit is currently recovered in 150 lb/in^2 gauge (1034 kPa gauge) steam; the rest is rejected in coolers or leaves in hot gases. The primary objective of the level 6 analysis is therefore maximization of the amount and the value of the recovered heat.

Pinch analysis showed that it would be possible to increase heat recovery to 150 lb/in^2 gauge (1034 kPa gauge) steam by more than 10 million Btu/h (2.93 MW, or nearly 31 percent) with realistic temperature driving forces for heat transfer. It would also be possible to increase the steam pressure to nearly 600 lb/in^2 gauge (4137 kPa gauge) with only a very small loss in the quantity of heat that could be recovered. This increase in pressure would allow the steam to be passed through a backpressure steam turbine for power generation, without losing its ability to satisfy heating demands at the 150 lb/in^2 gauge (1034 kPa gauge) level. However, it is doubtful that this would be economical in a retrofit situation.

Level 7 (equipment and pipework specifications)

Insufficient data were available to conduct this level of the analysis.

The ideas generated in the hierarchical review of the FCC unit were further evaluated by using conventional engineering methods. In some cases, further evolution of the concepts occurred, and the final form of the project differs somewhat from the original idea. Several of these concepts appear, in various guises, in the list of process improvement options shown in Table 12.1, together with ideas from the related studies of the crude unit and the sour water system. All the projects result in savings in raw materials and/or utilities (water, steam, or fuel) and provide environmental benefits. The potential process improvements include

- Elimination of the surplus water in the sour water system, removing a potential source of odors

- A 30 percent reduction in desalter brine flow

- Recovery of up to 7300 barrels (approximately 1160 m^3) per year of oil (with equivalent reduction in loading in the water treatment plant)

- Savings of more than 30 million Btu/h (8.8 MW) in fuel firing

- Recovery of an additional 20 million Btu/h (5.9 MW) in fuel gas

Conclusions

The use of a structured analysis framework, such as the levels of the decision hierarchy, can make the identification of process improvement options for reducing emissions much more rapid and efficient than conventional ad hoc methods. The approach is equally valid both for new designs and retrofit studies.

The hierarchical review approach described is especially applicable where technical options have not yet been defined. The procedure offers a systematic analysis of new or existing processes to identify underlying causes of emission problems and to develop effective means of eliminating or controlling them. This results in the definition of appropriate flow sheet structures and equipment selections. Other complementary process integration techniques, especially numerical optimization and pinch analysis, can then be used to fine-tune the design.

An important benefit of many of the process improvement ideas identified by this method is that they have the effect of lowering operating costs at the same time as reducing emissions. This was a major feature of the Amoco Yorktown study described in this chapter.

References

1. J. M. Douglas, "A Hierarchical Decision Procedure for Process Synthesis," *AIChEJ*, 31(3): 353–362, March 1985.
2. A. P. Rossiter, D. C. Woodcock, and J. M. Douglas, "A Hierarchical Decision Procedure for Retrofit Studies of Solids Systems," American Institute of Chemical Engineers national meeting, paper 38b, Miami Beach, November 1986.
3. N. M. Duty, "A Random Search Approach to Optimized Process Retrofits," American Institute of Chemical Engineers spring national meeting, Houston, April 1991.
4. A. P. Rossiter, H. D. Spriggs, and H. Klee, Jr., "Apply Process Integration to Waste Minimization," *Chem. Eng. Progress,* 89(1): 30–36, January 1993.
5. J. M. Douglas, *Conceptual Design of Chemical Processes,* McGraw-Hill, New York, 1988, pp. 108–109.
6. G. J. Papagiannes, "Select the Right Dryer," *Chem. Eng. Progress,* 88(12): 20–27, December 1992.
7. K. E. Nelson, "Use These Ideas to Cut Waste," *Hydroc. Proc.,* 69(3): 93–98, March 1990.
8. D. M. Haseltine, "Wastes: To Burn, or Not to Burn?" *Chem. Eng. Progress,* 88(7): 53–58, July 1992.
9. H. Klee, Jr. and M. Podar, "Pollution Prevention in Petroleum Refining: The Amoco/USEPA Project at Yorktown," American Institute of Chemical Engineers national meeting, paper 84b, Houston, April 1991.

Intelligent Process Design and Control for In-Plant Waste Minimization*

Y. L. Huang, Ph.D.
Department of Chemical Engineering
Wayne State University
Detroit, Michigan

Prof. L. T. Fan
Department of Chemical Engineering
Kansas State University
Manhattan, Kansas

Introduction

The process industry has been one of the major waste generators for many decades. It is highly desirable that wastes be minimized or even eliminated at their sources. One of the most effective means to accomplish this is through better process design and control, which is often difficult to perform through conventional algorithmic methods. This becomes obvious once the nature of *waste minimization* (WM) engineering is understood.

---

*Adapted with permission from "Artificial Intelligence for Waste Minimization in the Process Industry," *International Journal of Computers in Industry,* 22: 117–128, 1993.

- WM requires a multidisciplinary effort involving cooperative participation of experts from a multitude of fields such as engineering, chemistry, biology, fluid mechanics, mathematics, statistics, economics, and law.
- WM is heavily dependent on experience. Usually, the behavior of a process system generating wastes cannot be described readily through deterministic mathematical models; instead, it demands qualitative analysis and description.
- The available information pertaining to WM is frequently uncertain, imprecise, incomplete, and qualitative. It can be often expressed only in symbolic form.

When the conventional methods are inapplicable, it becomes more desirable or even necessary to rely on nonconventional techniques. Various methods of artificial intelligence (AI) are examples of such nonconventional techniques.[1,2]

Artificial Intelligence

AI is mainly concerned with mimicking human intelligence on a computer to deal with information and knowledge that are complex, symbolic, ambiguous, and uncertain. The major disciplines of AI include expert systems, fuzzy logic, and (artificial) neural networks.

An *expert* or *knowledge-based system* is a computer program utilizing knowledge and inference procedures to solve problems at an expert level, which are sufficiently difficult that significant human expertise is required.[2] Programs of this type are especially effective in acquiring both structured and unstructured symbolic knowledge, representing them structurally, and manipulating the resultant knowledge. *Fuzzy logic* is a logical system that is much closer in spirit to human thought and language than the conventional logical system; it uses rigorous mathematics to deal with imprecise information. Thus, somewhat paradoxically, fuzzy logic is a precise logical system for imprecise reasoning.[3,4] It is capable of dealing with imprecisely structured numerical knowledge. A *neural network* is a computing system comprising a number of highly interconnected processing elements, which processes information by its dynamic state response to external numerical inputs.[5] It effectively captures unstructured knowledge hidden in a pool of numerical data.

The implementation of WM is usually highly complicated. A wide range of knowledge has to be acquired, represented, and manipulated. Knowledge required for the tasks involved in WM can be structured or unstructured, numerical or symbolical, precise or imprecise, and certain or uncertain. For example, energy and mass balance relations and the laws of thermodynamics are expressed precisely in mathematical form and are computed numerically. Information on the disturbances and fluctuations experienced by a process is frequently imprecise and incomplete. Expert experience in designing, controlling, and diagnosing a process is usually expressed in symbolic form (IF-THEN rule). Moreover, in modeling a complex process generating a variety of wastes, the input and output data are often the only available information on the process. Obviously, the exploitation of knowledge of all types is the key to solving WM problems.

The adoption of sundry subdisciplines of AI in solving complex problems can be highly advantageous. We have developed four intelligent systems for WM: (1) a hybrid intelligent system for designing material and energy recovery processes for minimizing hazardous species and waste energy,[6] (2) a hierarchical expert system for minimizing waste generation through production scheduling in a batch chemical plant,[7] (3) a fuzzy expert system for minimizing cyanide-containing waste solutions in an electroplating plant,[8] and (4) a fuzzy expert system for fault detection in the operation of a hazardous waste incineration system.[9] This chapter introduces the first two intelligent systems.[6,7] Due to lack of space, the other two systems are not discussed further here. Readers requiring more information on these systems should consult the original papers.[8,9]

Hybrid Intelligent System for Designing Material and Energy Recovery Processes

A hybrid intelligent system, named HIDEN, has been built to synthesize *material and energy recovery processes* (MERPs) for minimizing hazardous or toxic waste streams and recovering waste energy in a process plant. This system adopts various subdisciplines of AI including the knowledge-based approach, fuzzy logic, and neural networks. It attempts to fully exploit knowledge at all levels, including the linguistic, conceptual, epistemological, logical, and physical levels.

System structure

System HIDEN consists of three subsystems: knowledge-based subsystem KBDEN, fuzzy logic subsystem FLDEN, and neural network subsystem NNDEN. The structure of system HIDEN is illustrated in Fig. 13.1.

Subsystem KBDEN is the core of HIDEN; its knowledge base is shown in Fig. 13.2. This subsystem accepts the process data pertaining to a MERP synthesis problem. The data are divided into two classes: those representing a steady-state operating condition, which are pre-

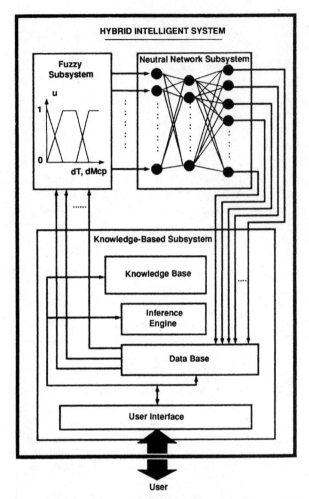

Figure 13.1   Structure of a hybrid intelligent system for synthesizing energy and material recovery processes.

cise, and those describing the dynamic operating condition, which are imprecise. The first class of data forms the bases on which the minimum number of process units and the minimum material or energy consumption are computed; the decisions on stream matching are made; and the complete structure of a process is constructed. The second class of data is transmitted to subsystem FLDEN.

Subsystems FLDEN and NNDEN are designed to provide a series of recommendations on critical stream matching derived from the second class of data. These data are first made fuzzy in subsystem FLDEN and then computed in subsystem NNDEN. Subsystem NNDEN generates a

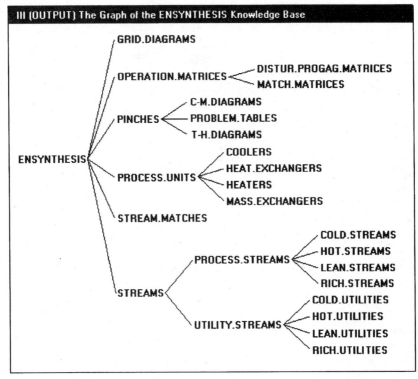

III (OUTPUT) The Graph of the ENSYNTHESIS Knowledge Base

GRID.DIAGRAMS

OPERATION.MATRICES
- DISTUR.PROGAG.MATRICES
- MATCH.MATRICES

PINCHES
- C-M.DIAGRAMS
- PROBLEM.TABLES
- T-H.DIAGRAMS

ENSYNTHESIS

PROCESS.UNITS
- COOLERS
- HEAT.EXCHANGERS
- HEATERS
- MASS.EXCHANGERS

STREAM.MATCHES

STREAMS
- PROCESS.STREAMS
  - COLD.STREAMS
  - HOT.STREAMS
  - LEAN.STREAMS
  - RICH.STREAMS
- UTILITY.STREAMS
  - COLD.UTILITIES
  - HOT.UTILITIES
  - LEAN.UTILITIES
  - RICH.UTILITIES

Figure 13.2 Structure of part of the knowledge base in subsystem KBDEN of hybrid intelligent system HIDEN.

multitude of the most superior and inferior stream-matching decisions in the light of synthesis criteria. All the stream-matching decisions derived are taken into account by subsystem KBDEN.

Subsystems KBDEN and FLDEN are built on a XEROX 1108 AI workstation using KEE (Knowledge Engineering Environment), a LISP-based object-oriented multiparadigm programming environment. Subsystem NNDEN resides in an IBM PC compatible using NeuralWare Professional II, a neural network development environment. These subsystems have been fully integrated into system HIDEN for designing a MERP. A large number of MERPs, each of which is cost-effective and highly controllable in terms of WM, have been synthesized. The concept of structural controllability of a process and an index of structural controllability have been fully developed.[10] To demonstrate the applicability of system HIDEN, two design problems—one for material recovery and the other for energy recovery—are solved.

Synthesis of a material recovery process

Mass exchangers are becoming increasingly important for material recovery processes in separating hazardous or toxic species from process streams.[11] The synthesis of a mass-exchanger network is illustrated by designing a process for sweetening coke oven gas (COG). It removes acidic impurities, of which hydrogen sulfide is the most significant. It is corrosive and contributes to SO_2 emission when the COG is consumed as a fuel.

The data pertinent to the synthesis problem stated above are listed in Table 13.1. Sour COG R_1 and tail gases R_2 from the "Claus unit" are rich streams. Aqueous ammonia L_1 and chilled methanol L_2 are lean streams. The concentration of hydrogen sulfide in each of the two rich streams should be reduced to a specific value through mass exchange with one or two lean streams. In the table the mass flow rate M and the source and target concentrations of hydrogen sulfide in the streams, $y^s_{P_R}$ and $y^t_{P_R}$ for the rich streams and $x^s_{P_L}$ and $x^t_{P_L}$ for the lean streams, represent the static operating conditions. The degree of disturbance D and the level of control precision C reflect the dynamic operating conditions. The process structure synthesized by system HIDEN is illustrated in Fig. 13.3. This solution (solution A in Table 13.2) is the optimal one compared with other feasible ones (solutions B through D) derived by exhaustive search, since it has the highest degree of structural controllability, while the capital and operating costs are the same as for the others.

Synthesis of an energy recovery process

The pertinent data for the energy recovery synthesis problem are given in Table 13.3, in which symbols H_1 and H_2 stand for the two hot streams whose temperatures should be reduced and C_1, C_2, and C_3 stand for the three cold streams whose temperatures should be increased. The heat capacity flow rate Mc_p and the stream source and

TABLE 13.1   Stream Data of the Material Recovery Synthesis Problem

Rich stream	Mass flow rate M, kg/s	Source conc. y^s_p	Target conc. y^t_p	Max. deviation of mass flow rate of component p, δM_p, kg/s	Max. allowable deviation of target conc. δy^t_p
R_1	0.9	0.0700	0.0003	0.0063 (moderate disturbance)	0.00002 (high precision)
R_2	0.1	0.0510	0.0001	0.0014 (severe disturbance)	0.00001 (high precision)

Lean stream	Mass flow rate M, kg/s	Source conc. x^s_p	Target conc. x^t_p	Max. deviation of mass flow rate of component p, δM_p, kg/s	Max. allowable deviation of target conc. $\delta x_p{}^t$
L_1	2.3	0.0006	0.0310	0.0084 (moderate disturbance)	0.0040 (moderate precision)
L_2	≤ 15	0.0002	0.0035	0.0001 (slight disturbance)	0.0009 (low precision)

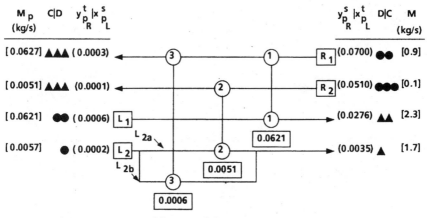

Figure 13.3   Solution structure of the material recovery process.

TABLE 13.2   Comparison of the Solutions for the Material Recovery Synthesis Problem

Solution	A (AI approach)	B	C	D
			←——Exhaustive search——→	
Consumption of mass separating agent, kg/s	0.0057	0.0007	0.0057	0.0007
Number of mass-transfer units	3	4	3	4
Degree of structural controllability	0.769	0.462	0.615	0.385

TABLE 13.3   Stream Data of the Energy Recovery Synthesis Problem

Hot stream	Heat capacity flow rate Mc_p, kW/°C	Source temp. T_s, °C	Target temp. T_t, °C	Max. deviation of heat load δQ, kW	Max. allowable deviation of target temp. δT_t, °C
H_1	13.29	204.4	65.6	150 (moderate disturbance)	2 (high precision)
H_2	16.62	248.9	121.1	250 (severe disturbance)	3 (high precision)

Cold stream	Heat capacity flow rate Mc_p, kW/°C	Source temp. T_s, °C	Target temp. T_t, °C	Max, deviation of heat load δQ, kW	Max. allowable deviation of target temp. δT_t, °C
C_1	13.03	93.3	204.4	150 (severe disturbance)	12 (low precision)
C_2	12.92	65.6	182.2	70 (slight disturbance)	11 (low precision)
C_3	11.40	37.7	204.4	75 (slight disturbance)	4 (high precision)

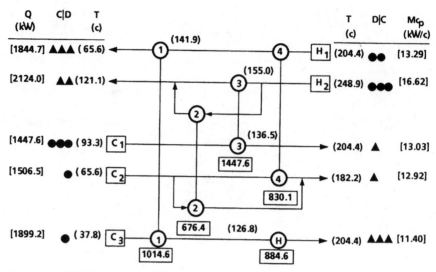

Figure 13.4   Solution structure of the energy recovery process.

TABLE 13.4   Comparison of the Solutions for the Energy Recovery Synthesis Problem

Solution	A (AI approach)	B	C	D	E	F
		$\leftarrow$		Exhaustive search		$\rightarrow$
Consumption of energy, kW	884.6	884.6	884.6	884.6	884.6	884.6
Number of heat-transfer units	5	5	5	5	5	5
Degree of structural controllability	0.857	0.700	0.571	0.667	0.571	0.286

target temperatures T_s and T_t represent the static operating conditions. The degree of disturbance and the level of control precision reflect the dynamic operating conditions. The process structure synthesized by system HIDEN is shown in Fig. 13.4. This solution (solution A in Table 13.4) is the optimal one when compared with other feasible ones (solutions B through F) derived by exhaustive search, since it has the highest degree of structural controllability, while the capital and operating costs are the same as for the others.

Hierarchical Expert System for Scheduling a Batch Process with WM

Batch processing is increasingly becoming popular for manufacturing value-added specialty chemicals. To simultaneously maximize the throughput and minimize the operating cost and waste generation, computerized production scheduling is almost always necessary.[12] Most of the earlier works resort to approaches based on mathematical programming techniques.

Various types of hazardous or toxic wastes or by-products are also generated in a batch chemical plant. It is highly desirable, therefore, that WM be considered an additional objective in minimizing the total cost of a plant. Unfortunately, much of the necessary information and available data related to waste generation are often incomplete, imprecise, and uncertain; therefore, expertise is crucial in decision making for scheduling. Consequently, the expert system is a viable alternative to conventional methods for scheduling.

Process description

The case studies given here are based on the main process of the typical batch chemical plant described in Rippin and Egli.[13] The plant, depicted in the lower part of Fig. 13.5, contains three reactors, two filters, and one dryer. In this multiproduct, multistage plant, four different products—EV_1, EV_2, F, and H—are to be manufactured from three groups of raw materials—E_1 through E_3, F_1 through F_3, and H_1 and H_2. The process involves the following:

1. *Different production sequences.* The manufacture of each product requires various types and numbers of process units. For instance, raw materials F_1 through F_3 and product E (which is also a raw material for product F) should be processed sequentially in reactors 2 and 3, filter 2, and dryer 1 with different residence times to yield final product F (Fig. 13.6).

2. *Generation of unstable intermediates.* Unstable intermediates are generated in manufacturing certain products. For example, in manufacturing product F, the intermediates from reactor 2 and those from filter 2 are unstable (Fig. 13.6). The unstable intermediates, designated by the symbol u in Fig. 13.5, should be withdrawn and charged into the succeeding process units as early as possible to prevent their decomposition; otherwise, waste may be generated and the environment may be contaminated. Hence, the time to withdraw the group of intermediates from these two reactors and one filter should be carefully determined in the light of the availability of succeeding process units. The prevention of decomposition of unstable intermediates is one of the main concerns of WM.

3. *Product interdependencies.* Some products, such as products E and F, are both intermediates and final products.

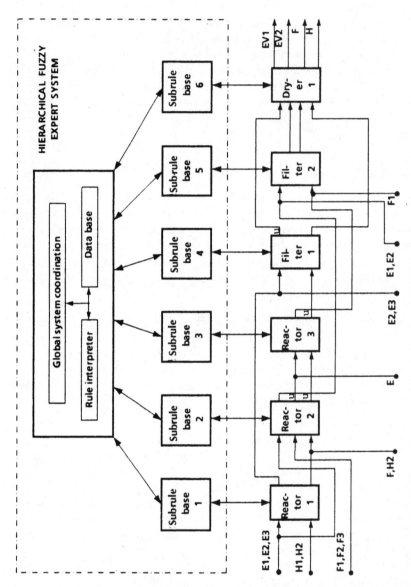

Figure 13.5 Batch process and the structure of the hierarchical expert system.

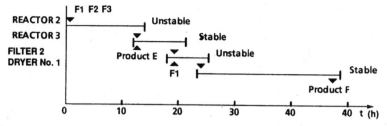

Figure 13.6 Processing sequence of product F.

4. *Changeover of a process unit.* A process unit usually needs to be cleaned after each batch and prepared for the next batch. The time needed for cleaning and preparation during the production period depends on the sequence chosen for the production variants. Also, statistical fluctuations in yield and the times required for various tasks may gradually accumulate, resulting in significant deviations from the planned schedule.

System structure

A hierarchical expert system is structured with a lower layer and an upper layer. The lower layer consists of a number of subrule bases, each of which contains a multitude of rules for scheduling a subprocess, essentially a specific process unit; the upper layer has a subrule base for coordinating all subprocesses. This is illustrated for the specific example of the scheduling system in the upper part of Fig. 13.5.

Two classes of knowledge—deep and shallow knowledge—are required for scheduling. Deep knowledge includes the first principles of reaction kinetics, thermodynamics, and mass and energy balances as well as the process models derived from these first principles. The operations of processes can be evaluated or predicted in the light of the models. For example, with the assumptions of perfect mixing and constant volume for a batch reactor, the reaction time in the reactor can be calculated as

$$t = C_{A0} \int_{x_{A0}}^{x_A} \frac{dx_A}{k(T)f(C_A)} \tag{13.1}$$

where t = reaction time
C_{A0} = initial concentration of reactant A
x_A = mole fraction of reactant A
$k(T)$ = reaction rate
$f(C_A)$ = fractional conversion

While deep knowledge can be expressed precisely in mathematical form without much difficulty, this is not the case with shallow knowledge. Shallow knowledge is usually expressed as heuristic rules or functions. Since the available information is imprecise, the heuristic rules are "fuzzy" in nature; such rules are termed fuzzy heuristic rules or simply fuzzy rules. An example is the following rule related to the selection of a group of materials for dryer 1:

IF   the relative instability of the groups of materials is low ($\bar{E}$) AND the relative priority is high (A)

THEN   charge the group of materials with the highest instability into dryer 1

This rule contains two fuzzy linguistic terms: low relative instability and high relative priority. These terms associated with fuzzy sets need to be defined. The rule base consists of six subrule bases, each of which is composed of a number of IF-THEN rules.[14]

Decision-making process

Decisions on scheduling are made in two stages. Each subsystem in the lower layer makes a locally optimal decision first. This is followed by the coordination of the subsystems by modifying the local decisions subject to the criterion of global optimality.

Local decision making.   Each rule in a subrule base at the lower layer is weighted first. The weight $\omega_{i\text{-}j}$ and the truth value $\tau_{i\text{-}j}$ collectively yield the weighted truth value $\tau_{i\text{-}j}^{\omega}$ for rule $i\text{-}j$, that is,

$$\tau_{i\text{-}j}^{\omega} = \tau_{i\text{-}j} \wedge \omega_{i\text{-}j} \tag{13.2}$$

The assignment of a $\omega_{i\text{-}j}$ to a rule is subjective; the weight is in the closed interval [0, 1]. The decision for each subsystem is made by the following manipulation:

$$D_i^{t_k} = \tau_{i\text{-}1}^{\omega} \vee \tau_{i\text{-}2}^{\omega} \vee \cdots \vee \tau_{i\text{-}n}^{\omega} \tag{13.3}$$

where $\tau_{i\text{-}j}$ = truth value of rule $i\text{-}j$, that is, rule j in subrule base i
$\tau_{i\text{-}j}^{\omega}$ = weighted truth value of rule $i\text{-}j$
$\omega_{i\text{-}j}$ = weight of rule $i\text{-}j$
$D_i^{t_k}$ = decision of subsystem i at time t_k

Then the general rule-firing mechanism is utilized. It is always desirable to fire the rules in parallel. This ensures that each rule has a commensurate opportunity to participate in decision making; thus, an undue dominance of any rule is avoided.

Coordination among local decisions. A set of globally optimal decisions is obtained by coordinating the local decisions according to the scheduling criteria, such as the minimum total idle time of process units in a day, meeting the production plan, and WM; a trade-off should always be made among these criteria. When a set of local decisions fails to satisfy the criteria, they are sent back with directions for changing the rule weights. Then new decisions are made and sent to the upper subsystem for evaluation.

PS-BATCH: A hierarchical expert system

Expert system PS-BATCH is developed for the scheduling of a batch chemical process.[15] This system is implemented on an IBM PC compatible using Personal Consultant Plus, an expert system shell. System PS-BATCH is of the question-conclusion type. It poses a set of questions regarding the operating conditions of each process unit. The user is required to respond to the questions.

Figure 13.7 illustrates some of the questions posed by the system and the responses from the user. These questions can be classified into two groups: yes/no and ambiguous types. For instance, the question "What is the group of raw materials with the highest priority to be charged into reactor 2?" is of the yes/no type, and the question "What is the accuracy of the changeover time of reactor 1?" is of the ambiguous type. The user needs to respond to them based on observations and estimation of the production.

Figure 13.8 lists the decisions made by the system for scheduling in a time period. This system gives a high priority to the prevention of decomposition of unstable intermediates. In other words, when more than one group of intermediates is waiting to be charged into a process unit, the most unstable group of intermediates is always selected first. If a group of unstable intermediates is almost ready to be withdrawn from a process unit, it has a higher priority to be charged to the next unit than another group of stable intermediates already waiting.

Conclusions

Artificial intelligence techniques are promising tools for waste minimization in the process industries. This has been amply demonstrated by a number of intelligent systems that have been successfully applied to applications of this type. The two intelligent systems described in this chapter illustrate this point.

COLSULTATION RECORD OF PA-BATCH:
A hierarchical fuzzy expert system for scheduling a multi-stage batch chemical plant

"High relative priority between two groups of materials to be charged into REACTOR 1:	Yes:80%"
"Group of materials with higher priority to be charged into REACTOR 1:	H"
"Accuracy of the changeover time of REACTOR 1:	Yes:40%"
"Preceding batch of the group of materials processed in REACTOR 1:	EV1"
"Product with the smallest quantity in stock:	H"
"Higher relative priority between three groups of materials to be charged into REACTOR 2:	Yes:80%"
"Group of materials with the highest priority to be charged into REACTOR 2:	F"
"Preceding batch of the group of materials processed	

Figure 13.7   Part of the record of human-machine communication of system PS-BATCH.

The major advantages of AI techniques are that imprecise information can be processed readily, engineering experience can be utilized fully, and effective solutions can be identified quickly. It is probable that AI techniques will play an increasingly important role in WM.

Acknowledgments

Although the research described in this chapter has been mainly funded by the Environmental Protection Agency under assistance agreement R–819653 to the Great Plains–Rocky Mountain Hazardous Substance Research Center for EPA regions 7 and 8 with headquarters at Kansas State University, it has not been subjected to the agency's peer and administrative review and, therefore, may not necessarily reflect the views of the agency, and no official endorsement should be inferred. The first author also gratefully acknowledges grants from the University

SCHEDULING DECISION MADE BY PS-BATCH: A hierarchical fuzzy expert system for scheduling a multi-stage batch chemical plant	
Group of materials selected to charge to REACTOR 1:	H
Group of materials selected to charge to REACTOR 2:	F
Group of materials selected to charge to REACTOR 3:	F
Group of materials selected to charge to FILTER 1:	H
Group of materials selected to charge to FILTER 2:	F
Group of materials selected to charge to DRYER:	E

Figure 13.8   Example of scheduling decision by system PS-BATCH.

Research Program and the Minority Faculty Research Program at Wayne State University.

References

1. E. Rich, *Artificial Intelligence,* Chap. 2, McGraw-Hill, New York, 1983.
2. D. A. Waterman and D. B. Lenat, *Building Expert Systems,* Addison-Wesley, Reading, MA, 1983.
3. L. A. Zadeh, "Foreword," in *Fuzzy Logic in Knowledge-Based Systems, Decision and Control,* M. M. Gupta and T. Yamakawa (eds.), North-Holland, Netherlands, 1988, pp. xiii–xiv.
4. L. A. Zadeh, "Fuzzy Logic and Approximate Reasoning," *Syntheses,* 30: 407–428, 1974.
5. D. E. Rumelhart, J. L. McClelland, and PDP Research Group, *Parallel Distributed Programming: Explorations in the Microstructure of Cognition,* vol. 1, MIT Press, Cambridge, MA, 1986.
6. Y. L. Huang and L. T. Fan, "HIDEN: A Hybrid Intelligent System for Synthesizing Highly Controllable Exchanger Networks: Implementation of a Distributed Strategy for Integrating Process Design and Control," *IE&C Res.,* 33: 1174–1187, 1994.
7. Y. L. Huang and L. T. Fan, "Hierarchical Fuzzy Expert System for Production Scheduling: A Case Study for Scheduling a Multiproduct Batch Chemical Process," presented at the American Institute of Chemical Engineers spring national meeting, paper 57e, Orlando, FL, March 18–22, 1990.
8. Y. L. Huang, G. Sundar, and L. T. Fan, "MIN-CYANIDE: An Expert System for Cyanide Waste Minimization in Electroplating Plants," *AIChE/Envir. Prog.,* 10: 89–95, 1991.
9. Y. L. Huang and L. T. Fan, "A Fuzzy-Logic-Based Approach to Building Efficient Fuzzy Rule-Based Expert Systems," *Comp. Chem. Eng.,* 17: 181–192, 1993.
10. Y. L. Huang and L. T. Fan, "Distributed Strategy for Integration of Process Design and Control: A Knowledge Engineering Approach to the Incorporation of Controllability into Exchanger Network Synthesis," *Comp. Chem. Eng.,* 16:497–522, 1992.

11. M. M. El-Halwagi, "Synthesis of Optimal Mass-Exchange Networks," Ph.D. dissertation, University of California, Los Angeles, 1990.
12. G. Grun, W. Fichtner, and W. Janiche, "Scheduling of a Multi-purpose Chemical Plant," presented in Third International Congress: Computers and Chemical Engineering, Pariers, 1981.
13. D. W. T. Rippin and U. M. Egli, "Short-Term Scheduling for Multiproduct Batch Chemical Plants," *Comp. Chem. Eng.*, 10: 302–325, 1986.
14. Y. L. Huang, "Integrated Process Design and Control of Chemical Processes via Artificial Intelligence Techniques: Application to Waste Minimization," Ph.D. dissertation, Chap. 9, Kansas State University, Manhattan, 1992.
15. Ibid., Chap. 11.

Knowledge-Based Design Approach for the Simultaneous Minimization of Waste Generation and Energy Consumption in a Petroleum Refinery

Y. L. Huang, Ph.D.
Department of Chemical Engineering
Wayne State University
Detroit, Michigan

Prof. Thomas F. Edgar
Department of Chemical Engineering
The University of Texas
Austin, Texas

Introduction

The petroleum refining industry has been a major source of hazardous or toxic waste streams containing various *volatile organic compounds* (VOCs), NO_x, and SO_x.[1,2] One of the main reasons is that refining processes have traditionally been designed without due consideration of in-plant *waste minimization* (WM).[3-6] In compliance with increasingly stringent environmental regulations, these processes and their respective flow sheets need to be reexamined and, if necessary, modified.

The waste streams being generated in oil refineries may differ appreciably in the types of hazardous or toxic species, forms (gas, liquid, slurry, solid, etc.), species concentrations, temperatures, and flow rates. All these should be taken into account in WM. Moreover, maintaining stable operating conditions is a major concern for any in-plant WM measures. This makes implementation of WM much more difficult than end-of-pipe waste treatment for which plant operation is less of a concern. In addition, the investment costs for WM should be kept as low as possible, which implies both economic and environmental objectives. To accomplish these objectives, the process modified for WM should be highly integrated. That is, the resulting process attempts to minimize the number of process units and materials and energy consumption.

It is conceivable that the integrated process will have a high degree of connectivity among its process units. Such a process, while in principle capable of minimizing waste with the lowest cost, may be extremely difficult to operate. If the process is not structurally controllable, the WM goals cannot be achieved. It is highly desirable, therefore, that a WM-oriented process be not only cost-effective but also highly controllable.[7] An additional requirement is the need to deal with a large amount of imprecise data. Moreover, the process should be modified based on not only first-principles modeling, but also design and operational experience. As indicated in Chaps. 9 and 13, a knowledge-based approach is preferred to traditional algorithmic approaches to treat these problem characteristics.

In this chapter, we focus on how to use a knowledge-based approach to design a process for the simultaneous minimization of both waste chemicals and energy consumption. The efficacy of the approach is demonstrated by its application to a phenol solvent extraction process in an oil refinery.

Simultaneous Minimization of Waste Chemicals and Energy Consumption

Effluent streams leaving refining processes usually contain various hazardous or toxic species. The species concentrations can often be reduced to permissible ranges by rearranging existing process units or integrating new separation and/or energy recovery processes into the existing plant. The most commonly used separation approach for this purpose utilizes mass-separating agents and is termed a *mass-exchange network* (MEN).[8] (See also Chap. 16.) A mass-exchange network performs unit operations such as extraction, absorption, and adsorption. In these operations, the species concentration in one process stream (the rich stream) can be reduced through mass exchange with another process stream (the lean stream) in a set of *mass-transfer units* (MTUs). Due to temperature differences, waste-species-containing streams need to be preheated or precooled to meet the temperature specifications of each individual unit operation. To minimize energy consumption, a *heat-exchange network* (HEN) can be employed. A process containing a MEN and an HEN is termed a *combined-exchange network* (CEN).[9] It is capable of reducing the quantity and toxicity of waste streams as well as energy consumption. However, the CEN may be much more difficult to operate because of the existence of numerous thermal and mass interactions among process streams. Although the operability analysis of an HEN has been conducted,[10,11] there have been no available approaches for the operability enhancement on a CEN, which is much more complicated than an HEN.

Knowledge-Based Design Approach

A knowledge-based approach is proposed to synthesize a cost-effective and highly controllable CEN for the simultaneous minimization of both waste generation and energy consumption. In this development, the domain knowledge—both deep knowledge and shallow knowledge—should be fully utilized. Deep knowledge, i.e., the first principles, uncovers the fundamental cause-effect relationships, while shallow knowledge relies on engineering experience and other heuristics.[12] All the knowledge should be appropriately represented and manipulated.

Knowledge representation

First-principles knowledge. The first-principles knowledge required for the design mainly includes mass and energy balance relations and the first and second laws of thermodynamics. They are usually represented as mathematical formulas. This is illustrated in App. A, which shows the relationships for mass transfer through a mass-transfer unit and heat transfer through a heat-transfer unit.

Heuristic rules. Shallow knowledge is usually expressed as rules and mathematical expressions. For instance, one such rule might be:

IF   the phenol concentration of a stream fluctuates severely

and   the phenol concentration of another stream should be controlled
          very precisely

THEN   these two streams should not be matched in an MTU.

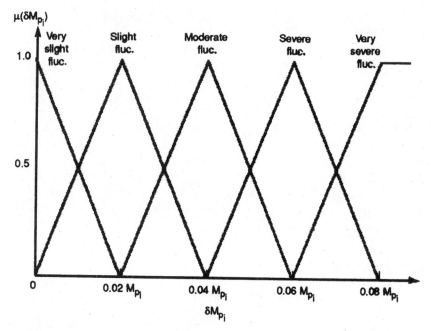

Figure 14.1 Quantitative representation of the fuzzy linguistic terms *very severe, severe, moderate, slight,* and *very slight fluctuations.*

In this rule, the linguistic terms *severely* and *very precisely* are fuzzy. It is extremely difficult to identify a sharp boundary between linguistic concepts such as severe and moderate fluctuations, and very precise and precise control. This is circumvented by resorting to fuzzy set theory.[13] Fuzzy sets A_1, A_2, A_3, A_4, and A_5 can be defined to represent the fuzzy concepts of very severe, severe, moderate, slight, and very slight fluctuations, respectively (Fig. 14.1). Fuzzy sets B_1, B_2, and B_3 can be defined to represent the fuzzy concepts of high, moderate, and low control precision, respectively. We consider only two match modes—match or no match—which implies very clear decisions on stream matching. The match modes are not fuzzy and thus can be expressed by ordinary sets C_1 and C_2. However, we still consider them as fuzzy sets, for consistency, since an ordinary set is a special case of a fuzzy set. The rule can be mathematically expressed in a fuzzy logic form for further computation:

$$A_1 \wedge B_1 \rightarrow C_2 \qquad (14.1)$$

A large number of fuzzy heuristic rules have been developed for CEN synthesis.[9] These rules are distributed in a hierarchically structured knowledge base and classified into sets for (1) measuring the toxicity of species, (2) selecting a method for separating hazardous or toxic species, (3) determining a separation sequence of stream components, (4) evaluating the feasibility of recycling hazardous or toxic species, (5) selecting the strategies for recovering waste energy, (6) implementing minimum capital and operating costs, (7) enhancing structural controllability, (8) modifying a synthesized process structure, and (9) making a tradeoff among capital cost, operating cost, WM, and structural controllability.

Structural controllability. The degree of structural controllability of a process reaches a maximum when the occurrences of undesirable disturbance propagations and their severities are at a minimum, thereby minimizing the possibility of waste generation due to the process structure. Thus, structural controllability can be assessed by examining the patterns of disturbance propagation throughout the process. This can be evaluated quantitatively, as discussed in App. B. A more detailed explanation is given in Huang and Fan.[14]

In general, the least controllable processes have the maximum generation of waste, which is definitely undesirable. The most controllable processes have the minimum generation of waste, but these are also generally undesirable due to extremely high capital and operating costs. In fact, it is not usually feasible to completely eliminate waste generation in a process plant. Our goal is to design a highly controllable process which is also economical to build and operate.

Knowledge manipulation

Computation of fuzzy rules. First-principles knowledge can be readily treated numerically. Heuristic knowledge, however, should be manipulated numerically as well as symbolically. This can be accomplished by resorting to fuzzy logic. Each heuristic rule needs to be expressed in a fuzzy logic form. The truth values of the fuzzy linguistic variables in the rule should be evaluated numerically. A decision is made according to the rule-firing algorithm, the fuzzy min-max algorithm.[13] In the evaluation of rules encoded in a knowledge base, the algorithm is employed at each rule level to select one rule for activation. The procedure is discussed further in App. C.

Synthesis procedure. The knowledge formalized by the approach described in the preceding subsection should be organized systematically in order to identify the most desirable solution effectively. In this procedure, the synthesis task is decomposed into three subtasks: (1) the minimization of waste stream generation by synthesizing a MEN, that is, to minimize the concentrations of hazardous or toxic species in effluent streams; (2) the minimization of energy consumption by synthesizing a HEN, that is, to maximize the use of energy available in the plant; and (3) the formation of a CEN by combining the MEN and the HEN.

A stagewise procedure for CEN synthesis is developed for this purpose. This procedure consists of five major modules: preanalysis, structure invention, structure modification, WM enhancement, and stream matching. The major function of the preanalysis module includes the estimation of both capital and operating costs and making decisions on stream matching which should lead to the least waste generation. The costs are estimated by pinch analysis.[15] (See also Part 2 of this book.) The structure invention module is designed for making decisions on the selection and placement of process units to match pairs of process streams. Each decision affects WM capability, which is assessed through evaluation of the index of structural controllability. Sets of rules for reducing the total cost and improving WM are applied to ensure identification of the most desirable solution. This module needs to activate both the WM enhancement module and the stream-matching module repeatedly. The resulting CEN flow sheet is examined by the structure modification module. Often when a number of separately synthesized subsystems in the structure invention module are combined to form a complete process, two undesirable situations may arise: The total number of process units exceeds the minimum requirement, and WM capability may deteriorate due to newly introduced intense disturbance propagations. Hence, tradeoffs need to be made among the number of process units, energy and material consumption, WM capability, and structural controllability. The tradeoff is usually based on engineering judgment. A detailed description of the functions of each module can be found in Edgar and Huang.[6]

Phenol Waste Minimization with Minimum Energy Consumption

The knowledge-based approach proposed here has been applied to synthesizing a CEN for minimizing phenol waste with minimum energy consumption in an oil refinery. Phenolics are considered one of the major organic toxic species that should be minimized. Phenol-containing waste streams are generated from a number of processes in an oil refinery, such as phenol solvent extraction and catalytic cracking.[2] The former process is considered here.

Process description

The phenol solvent extraction process is illustrated in Fig. 14.2. In this process, the phenol-containing streams come from three major process units: the raffinate tower, water/phenol tower, and extract stripper.[2] These streams are mixed first and then enter an absorber, where the

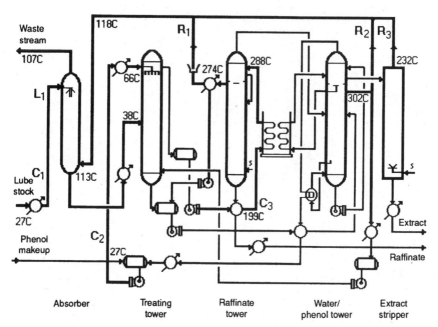

Figure 14.2   Phenol solvent extraction process.

heated lubricating oil stock absorbs phenol from the mixed stream. The waste stream leaving the process for a waste treatment plant usually contains excess phenol far beyond a tolerable limit (≤ 0.2 percent).

Note that the temperatures, concentrations, and flow rates of these three streams are different; the fluctuations of these variables are stream-independent. Mixing these streams is clearly inefficient thermodynamically and thus undesirable from the standpoint of phenol waste minimization. This suggests that a process capable of simultaneously minimizing both waste chemicals and energy needs to be synthesized.

Process preanalysis

The process data for the synthesis are given in Table 14.1. Three streams from the raffinate tower, water/phenol tower, and extract stripper are designated rich streams R_1, R_2, and R_3, respectively; they are rich in terms of phenol waste species. A mass balance computation indicates that lubricating oil alone is incapable of reducing the phenol concentration in the mixed stream to a tolerable limit, especially when severe disturbances appear at the inlet of the stream. A family of solvents could be used for recovering phenol. In this case, however, a *mass-separating agent* (MSA)—activated carbon—is used. The minimum quantity of activated carbon is expected to minimize the total cost. The lubricating oil and activated carbon are lean streams L_1 and

TABLE 14.1   Specification of Process Streams for the CEN Synthesis

Rich or lean stream	Mass flow rate M (kg/s)	Source concentration y_p^s or x_p^s	Target concentration y_p^t or x_p^t	Maximum deviation of mass flow rate of component p, δM_p (kg/s)	Maximum allowable deviation of target concentration δy_p^t or δx_p^t
R_1	1.95	0.15	0.002	0.0114 (moderate disturbance)	0.0001 (high precision)
R_2	0.68	0.10	0.001	0.0055 (severe disturbance)	0.0001 (high precision)
R_3	1.45	0.07	0.0015	0.0090 (severe disturbance)	0.0003 (moderate precision)
L_1	5.45	0.002	0.08	0.0118 (moderate disturbance)	0.01 (low precision)
L_2	≤5.00	0.0007	≤0.01	0.0005 (slight disturbance)	0.0005 (low precision)

Hot or cold stream	Heat capacity flow rate Mc_p (kW/°C)	Source temperature T_s (°C)	Target temperature T_t (°C)	Maximum deviation of heat load δQ (kW)	Maximum allowable deviation of target temperature δT_t (°C)
$H_1 (R_1)$	3.80	274	118	21.1 (slight disturbance)	3 (high precision)
$H_2 (R_1)$	3.26	107	600	6.1 (slight disturbance)	5 (moderate precision)
$H_3 (R_2)$	1.37	302	118	19.5 (severe disturbance)	3 (high precision)
$H_4 (R_2)$	1.08	107	38	3.5 (moderate disturbance)	8 (low precision)
$H_5 (R_3)$	2.43	232	118	23.2 (severe disturbance)	3 (high precision)
$H_6 (R_3)$	1.82	107	38	6.3 (moderate disturbance)	10 (low precision)
$H_7 (L_1)$	11.39	113	38	40.1 (moderate disturbance)	3 (high precision)
$C_1 (L_1)$	9.79	27	113	42.2 (moderate disturbance)	6 (moderate precision)
C_2	10.63	27	66	9.0 (low disturbance)	6 (moderate precision)
C_3	8.62	199	288	24.8 (low disturbance)	4 (high precision)

L_2, respectively. Note that the mass flow rate and phenol concentration at the outlet of stream L_2 in the table are the upper limits. Hot streams H_1 through H_7 and cold stream C_1 are identified by examining the temperature specification of the segments of rich streams R_1 through R_3 and lean stream L_1. Moreover, cold streams C_2 and C_3 are available in the original process. Stream C_2 is the pure phenol stream from phenol storage to the treating tower. Stream C_3 is the intermediate stream from the treating tower to the raffinate tower. The energy required for heating these two cold streams can be obtained from hot rich streams; in this way waste energy can be fully utilized. The intensity of disturbances appearing at the inlets of streams and the requirement of control precision of phenol concentrations at the outlets of streams are computed based on incomplete and imprecise information.

Synthesis of a combined-exchange network

A combined-exchange network is developed by first synthesizing a MEN, then generating an HEN, and finally combining these two exchange networks into a complete process system. For the MEN synthesis, the pinch point is located at the lean end of the composite curve in a concentration–mass load of phenol species diagram. The minimum number of process units to be utilized is five, and the optimal solution is shown in Fig. 14.3. It contains three absorbers (the three mass exchangers on stream L_1) and two adsorbers (the two mass exchangers on stream L_2). The minimum mass of phenol transferred to the activated carbon, determined from the pinch analysis, is 0.0309 kg/s. The sum of the two mass loads on stream L_2 in Fig. 14.3 is 0.0102 + 0.0207 = 0.0309 kg/s, so the target is achieved, and consumption of activated carbon is minimized in this design. Moreover, it is highly controllable, as evaluated by the index of structural controllability (see App. B).

By differentiating stream segments involved in the MEN, an eight-stream HEN synthesis problem is identified. The pinch point is located at 204.4°C for hot streams and 198.9°C for cold streams. The minimum energy requirement is 302.2 kW of steam and 417.8 kW of cooling water. This HEN requires at least 13 HTUs for maximum energy recovery. The resultant HEN is given in Fig. 14.4.

The CEN structure is illustrated in Fig. 14.5. This process is capable of maintaining the concentration of phenol species in the effluent streams in the prespecified limit (≤0.2 percent), even if severe disturbances and fluctuations occur during process operation. This process modification requires no further treatment of phenol-waste-containing streams. Meanwhile, energy contained in the waste streams is recovered to the maximum extent to heat two intermediate streams C_2 and C_3. Compared to the original process, the new process can save up to 85

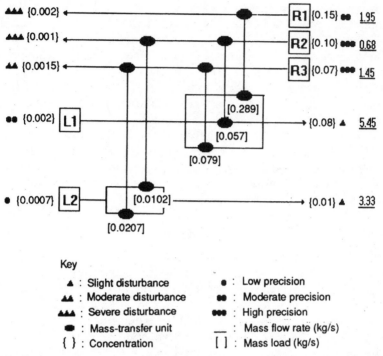

Figure 14.3   Flow sheet of the mass-exchange network for minimizing phenol waste.

percent of hot utility and 58 percent of cold utility. The overall degree of structural controllability is 0.82, based on the index of structural controllability. Thus, the CEN is highly controllable. Although modification of the process requires some new investment in HENs and MENs, this increase in process capital cost will reduce the capital cost of wastewater treatment facilities in the same refinery.

Conclusions

In-plant WM is becoming one of the key issues in the development of the petroleum refining industry. To minimize the overall cost of preventing the generation of waste streams, the energy available in a refinery should be utilized to the maximum extent. This implies the simultaneous minimization of both waste chemical generation and energy consumption. In this chapter, a knowledge-based design approach is developed for synthesizing a cost-effective and highly controllable combined-exchange network. The advantage of the approach over conventional algorithmic approaches is that it can manipulate both the first-principles knowledge and heuristic knowledge and can deal with both precise and imprecise information pertaining to the

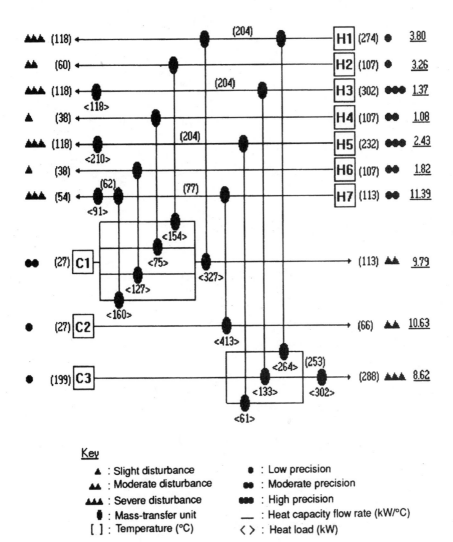

Figure 14.4 Flow sheet of the heat-exchange network for minimizing energy consumption in phenol waste minimization.

design problem. This approach should be useful for tackling other WM problems in an oil refinery, when additional knowledge required for problem solving is added to the knowledge base.

Acknowledgment

The first author gratefully acknowledges grants from the University Research Program and the Minority Faculty Research Program at Wayne State University.

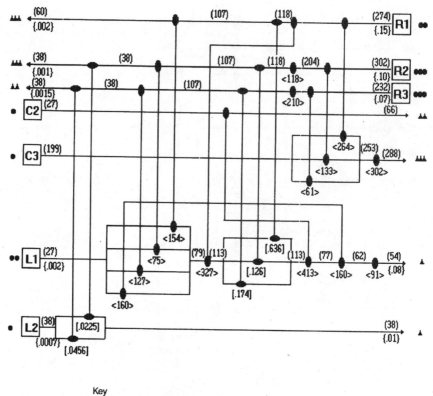

Key

▲	: Slight disturbance	●	: Low precision
▲▲	: Moderate disturbance	●●	: Moderate precision
▲▲▲	: Severe disturbance	●●●	: High precision
●	: Mass-transfer unit	▌	: Heat-transfer unit
{ }	: Concentration	( )	: Temperature (°C)
[ ]	: Mass load (kg/s)	< >	: Heat load (kW)

Figure 14.5 Flow sheet of the combined-exchange network for the simultaneous mini-mization of phenol waste and energy consumption.

Appendix A   Mass-Transfer and Heat-Transfer Relationships

The equilibrium solubility relation for a key component between a rich stream and a lean stream through a mass-transfer unit (MTU) is often approximated as the following linear relation:

$$C_{P_R} = aC_{P_L} + b \qquad (A.1)$$

where C_{P_R} = concentration of key component p in rich stream
C_{P_L} = concentration of key component p in lean stream
a, b = constants

The equation describing an operating line in an MTU is

$$\dot{M}_R (C_R^s - C_R^t) = \dot{M}_L (C_L^t - C_L^s) \qquad \text{(A.2)}$$

or it can simply be written in terms of the mass load of the key component p denoted $\dot{M}_p$, for each stream as

$$\dot{M}_{p_R} = \dot{M}_{p_L} \qquad \text{(A.3)}$$

where C_R^s = source concentration of rich stream
$\quad C_R^t$ = target concentration of rich stream
$\quad C_L^s$ = source concentration of lean stream
$\quad C_L^t$ = target concentration of lean stream
$\quad \dot{M}_R$ = normal mass flow rate of rich stream
$\quad \dot{M}_L$ = normal mass flow rate of lean stream
$\quad \dot{M}_{p_R}$ = mass load of key component p of rich stream
$\quad \dot{M}_{p_L}$ = mass load of key component p of lean stream

In an HEN, an equilibrium relation exists between a hot stream and a cold stream through a heat-transfer unit. This equilibrium relation can be written as

$$T_H = T_C \qquad \text{(A.4)}$$

where T_H = temperature of hot stream H
$\quad T_C$ = temperature of cold stream C

The equation describing an operating line in an HTU is as follows:

$$\dot{M}_{Cp_H} (T_H^s - T_H^t) = \dot{M}_{Cp_C} (T_C^t - T_C^s) \qquad \text{(A.5)}$$

This can simply be written in terms of the heat load $\dot{Q}$ for each stream as

$$\dot{Q}_H = \dot{Q}_C \qquad \text{(A.6)}$$

where T_H^s = source temperature of hot stream
$\quad T_H^t$ = target temperature of hot stream
$\quad T_C^s$ = source temperature of cold stream
$\quad T_C^t$ = target temperature of cold stream
$\quad \dot{M}_{Cp_H}$ = normal heat capacity flow rate of hot stream
$\quad \dot{M}_{Cp_C}$ = normal heat capacity flow rate of cold stream
$\quad \dot{Q}_H$ = heat load of hot stream
$\quad \dot{Q}_C$ = heat load of cold stream

Appendix B   Quantification of Structural Controllability

It is convenient to define disturbance vector D, control precision vector C, and disturbance propagation matrix P for a process having N streams. Vector D comprises all existing disturbances which are detrimental to WM. Each element in D represents the intensity of a disturbance exerted at the inlet of a specific stream. Vector C specifies the level of control precision required for each output variable. Each element in C represents the control precision required at the outlet of a stream. Matrix P lists all disturbance propagations in the process. The value of each element in P corresponds to the intensity of the disturbance propagation. Each element in D, C, and P is quantitatively evaluated by fuzzy set theory.[14]

To facilitate the evaluation of structural controllability in terms of WM and to compare various feasible solutions, an index I_{sc} is created. This has the following general form:

$$I_{sc} = \frac{E_{tot,max} - E_{tot}}{E_{tot,max} - E_{tot,min}} \tag{B.1}$$

where $E_{tot,max}$, $E_{tot,min}$, and E_{tot} are functions of vectors D and C and matrix P. The explicit form of each term, which was derived for MEN and HEN synthesis, can be found in Huang and Fan.[14] And $E_{tot,max}$ and $E_{tot,min}$ characterize a process with maximum and minimum disturbance propagations, respectively. Consequently, the former is the least controllable with the maximum generation of waste, which is definitely undesirable; the latter is the most controllable with the minimum generation of waste but is also undesirable due to extremely high capital and operating costs. In fact, to completely eliminate waste generation in a process plant is usually impossible. Our goal is to design a highly controllable process which is characterized by the term E_{tot}.

Appendix C   The Min-Max Algorithm

The min-max algorithm is illustrated by the following procedure for making an intermediate decision at the jth rule level where there are M rules. These rules are evaluated as

$$\mu_j(x) = \max\{\min\{\mu_{i1}(x), \mu_{i2}(x), ..., \mu_{ik}(x), ..., \mu_{iN}(x)\}\} \quad j \in J, \ i \in I \tag{C.1}$$

where $\mu_{ik}(x)$ = membership function of variable x in fuzzy set k representing kth antecedent of ith rule at jth level.

$\mu_j(x)$ = membership function of variable x in fuzzy set pertaining to rule selected to be fired at jth level

x = variable

For instance, the antecedent of the rule in the preceding subsection can be represented by the truth-value expression

$$\tau_i = \min\{\mu_{A1}(x), \mu_{B1}(x)\} \qquad \text{(C.2)}$$

where τ_i = truth value of rule i

$\mu_{A1}(x)$ = membership function of variable x in fuzzy set A_1

$\mu_{B1}(x)$ = membership function of variable x in fuzzy set B_1

x = phenol concentration

Operation min yields a set of M truth values through evaluation of the membership functions of all the related rules. Then a single rule is selected by performing operation max:

$$\tau = \max\{\tau_1, \tau_2, ..., \tau_M\} \qquad \text{(C.3)}$$

The selected rule is activated or fired. The same operation is repeated at the succeeding level based on the information received from the preceding level.

References

1. J. H. Gary and G. E. Handwerk, *Petroleum Refining: Technology and Economics,* 3d ed., Marcel Dekker, New York, 1994.
2. W. L. Nelson, *Petroleum Refinery Engineering,* 5th ed., McGraw-Hill, New York, 1969, pp. 360–362.
3. H. M. Dey, *Hazardous Waste Minimization: Potential and Problems for Texas Refineries,* M.P.A. Report, University of Texas at Austin, 1990.
4. D. T. Cindric, B. Klein, A. R. Gentry, and H. M. Gomaa, "Reduce Crude Unit Pollution with These Technologies," *Hydroc. Proc.,* 8: 45–54, August 1993.
5. A. P. Rossiter, H. Klee, Jr., and R. D. Sung, "Process Integration Techniques for Pollution Prevention," presented at the American Chemical Society IE&C Special Symposium, Atlanta, September 21–23, 1992.
6. T. F. Edgar and Y. L. Huang, "An Artificial Intelligence Approach to the Synthesis of a Process for Waste Minimization," in *Emerging Technologies in Hazardous Waste Management,* chap. 6, D. W. Tedder and F. G. Pohland (eds.), American Chemical Society, New York, 1993.
7. Y. L. Huang and L. T. Fan, "Artificial Intelligence for Waste Minimization in the Process Industry," *Int. J. Comp. Indus.,* 22: 117–128, 1993.
8. M. M. El-Halwagi and V. Manousiouthakis, "Synthesis of Mass Exchange Networks," *AIChE J.,* 35: 1233–1244, 1989.
9. Y. L. Huang, "Integrated Process Design and Control of Chemical Processes via Artificial Intelligence," Ph.D. dissertation, chap. 8, Kansas State University, Manhattan, 1992.
10. I. E. Grossmann and M. Morari, "Operability, Resiliency and Flexibility—Process Design Objectives for a Changing World," presented at the Second International

Conference on the Foundations of Computer-Aided Process Design, Snowmass, CO, June 1983.

11. R. D. Colberg and M. Morari, "A Flexibility Index Target for Heat Exchanger Networks," presented at the American Institute of Chemical Engineers annual meeting, New York, November 15–20, 1987.

12. A. Barr and E. A. Feigenbaum, *The Handbook of Artificial Intelligence,* vol. 1, HeurisTech Press, Stanford, CA, 1982, pp. 3–11.

13. H. J. Zimmermann, *Fuzzy Set Theory—and Its Application,* Kluwer-Nijhoff, Hingham, MA, 1985.

14. Y. L. Huang and L. T. Fan, "A Distributed Strategy for Integration of Process Design and Control: A Knowledge Engineering Approach for Incorporation of Controllability into Process Design," *Comp. & Chem. Eng.,* 16: 497–522, 1992.

15. B. Linnhoff, D. W. Townsend, D. Boland, G. F. Hewitt, B. E. A. Thomas, A. R. Guy, R. H. Marsland, J. R. Flower, J. C. Hill, J. A. Turner, and D. A. Reay, *User Guide on Process Integration for the Efficient Use of Energy,* The Institute of Chemical Engineering, London, 1982.

Numerical Optimization Approaches

15

Introduction to Numerical Optimization Approaches to Pollution Prevention

Mahmoud M. El-Halwagi, Ph.D.
Department of Chemical Engineering
Auburn University
Auburn, Alabama

Introduction

Over the past decade, several industries have been actively developing and implementing various designs for pollution prevention. Most of these designs have been devised to meet the specific characteristics of the process of interest. This approach has resulted in major hurdles that restrict the transfer of technology and expertise from one pollution prevention situation to another. In the context of a given pollution prevention task, the engineer is typically confronted with design decisions that require making choices from a vast number of options. These decisions involve the selection of the type of pollution prevention technology, system components, interconnection of units, and operating conditions. In many cases, one can envision lots of alternatives that are much too numerous to list. Typically, an engineer charged with the responsibility of designing a waste management system examines only a few process options, based on experience and corporate preference. Consequently, the designer selects the alternative with the most promising economic potential and designates it as the "optimum" solution. By assessing only a limited number of alternatives, one may easily miss the true optimum design or even get trapped in a region that is significantly different from the optimal one. In addition, the likelihood of generating innovative designs is severely reduced owing to the exclusive dependence on previous experience. These limitations call for the incorporation of numerical approaches into the design framework. Mathematical programming or optimization is the most general tool for doing this. In some cases, though, simple graphical methods can be used very effectively. Both of these approaches are discussed in Part 4 of the book.

Mathematical programming techniques have emerged in recent years as instrumental tools in systematizing the design of waste management systems. This trend has been accelerated by virtue of two factors—the tremendous growth in the use of computers and the recent advances in developing large-scale optimization algorithms. Graphical methods, on the other hand, have the advantage of providing a clear visual representation of results, which many people find helpful. Graphical methods can also be carried out without sophisticated algorithms or high-powered computers, making them accessible to virtually every engineer.

The overall goal of this chapter is to provide an introductory survey of the various optimization approaches that have been developed in recent years to tackle pollution prevention problems. Implicit in this goal is the additional objective of outlining the usefulness and limitation of these techniques.

What Is Mathematical Programming?

Mathematical programming (or optimization) involves the task of identifying the value of an n-dimensional vector x that minimizes (or maximizes) a certain quantity called the *objective function* $f(x)$ subject to E equality constraints and m inequality constraints, $h(x)$ and $g(x)$, respectively.

Mathematically, the problem can be stated as follows:

$$\min_x f(x)$$

subject to

$$h(x) = 0$$

$$g(x) \leq 0$$

where $x^T = [x_1\, x_2 \ldots x_n]$, $h^T(x) = [h_1(x)\, h_2(x) \ldots h_E(x)]$, $g^T(x) = [g_1(x)\, g_2(x) \ldots g_E(x)]$ and T is the transpose of a vector.

This definition of an optimization program is rich enough to embed any pollution prevention problem. For instance, the objective may be to minimize the cost of the system or the amount of waste. Similarly, the purpose of an optimization program may be to maximize the recovery of generated wastes or the economic potential of the process. Examples of equality constraints include material and energy balances, process modeling equations, and thermodynamic requirements. On the other hand, the nature of inequality constraints may be environmental (e.g., concentration of certain pollutants should be below specific levels), technical (e.g., pressure, temperature, or flow rate should not exceed some given values), and thermodynamic (e.g., driving force for mass, heat, or momentum transfer should be positive).

An optimization problem in which the objective function as well as all the constraints are linear is called a *linear program* (LP); otherwise, it is referred to as a *nonlinear program* (NLP). The nature of optimization variables also contributes to the classification of optimization programs. An optimization formulation that contains continuous (real) variables (e.g., pressure, temperature, flow rate) as well as integer variables (for example, 0, 1, 2, ...) is called a *mixed-integer program* (MIP). Depending on the linearity characteristics of the problem, MIPs can be further classified into *mixed-integer linear programs* (MILPs) and *mixed-integer nonlinear programs* (MINLPs). A particularly useful class of integer variables is the 0/1. These binary integer variables can be used to model logical events and decisions. For instance, a 0/1 variable may be designated to assume the value of 1 when an event occurs (e.g., a certain pollution prevention technology or unit is used) and 0 when that event does not occur.

The principles of optimization theory and algorithms are covered in various books.[1-4] Furthermore, several software packages are now commercially available. Many of these computer codes have been surveyed by Edgar and Himmelblau[1] (see pages 281, 335, 362, and 415). It is worth pointing out that most optimization software can efficiently obtain the global solution of LPs and MILPs. However, these commercial packages are not guaranteed to identify the global solution of nonconvex NLPs and MINLPs. Recently, significant research has been undertaken to develop effective techniques for the global solution of nonconvex NLPs and MINLPs.[5-8] Within the next few years, these endeavors may indeed culminate into practical procedures for globally solving general classes of NLPs and MINLPs.

Strategies of Using Mathematical Programming in Pollution Prevention

As stated earlier, pollution prevention is a highly combinatorial task. Numerical approaches can be employed to determine solutions while circumventing the combinatorial problem. Such numerical approaches can be roughly classified into two major strategies: decomposition and structural representation. The *decomposition* (or targeting) approach is based on tackling the pollution prevention task via a sequence of stages. Within each stage, a design target can be identified and employed in subsequent stages. For instance, consider the objective of minimizing the *total annualized cost* (TAC) of a waste management system (i.e., minimize TAC = annual operating cost + annualized fixed cost). This objective can be decomposed into two targets: minimize operating cost, then minimize annualized fixed cost. Hence, in the first stage, the minimum operating cost can be determined. In the second stage, the minimum fixed cost that satisfies the operating-cost target can then be obtained. Owing to the interaction between the two design targets, one must trade off these objectives through an iterative process. For instance, the designer can relax the requirement of minimum operating cost and pass the new solution to the second design stage, thereby obtaining a new TAC. The various TACs obtained iteratively are then compared to identify the design with the lowest TAC. It is noteworthy that this approach is not guaranteed to yield the minimum TAC of the system since the decomposed problems do not correspond exactly to the original problem. Nonetheless, the targeting approach offers two main advantages. First, within each stage, the problem dimensionality is reduced to a manageable size, thereby avoiding the combinatorial problems. Second, the decomposed problems typically feature mathematical properties (e.g., linearity) that render them amenable for global solution.

The second mathematical programming strategy that can be used in pollution prevention is the *structural-representation* approach. This technique models the system via a framework that embeds all potential configurations of interest. These configurations may include different types of pollution prevention technologies as well as their interconnections. Within the same structural representation, one can also embed all possible variations in design and operating variables. This approach is typically modeled as an MINLP whose solution extracts the optimal design from the potential candidates. While this approach is more robust than decomposition, its success is strongly intertwined with two challenging facets. First, the system representation should embed as many potential alternatives as possible. Failure to incorporate certain configurations may result in suboptimal solutions. The second limitation of the structural-representation approach stems from the inability of commercial optimization algorithms to guarantee the global solution of general MINLPs.

The decomposition and structural-representation approaches have been successfully employed in tackling pollution prevention tasks.

Several applications will be discussed in the remainder of this chapter as well as in Chaps. 16 to 18.

Classification of Numerical Approaches to Pollution Prevention

A distinct feature of employing mathematical tools in developing systematic techniques for pollution prevention is the ability of the developed procedures to cover a wide range of applications. Hence, one may characterize these approaches according to the scope of potential applications. In this context, the following categories can be used to classify the key milestones in the development of numerical techniques for the systematic design of pollution prevention systems:

1. Design of separation systems for the recycle and reuse of pollutants

2. Synthesis of environmentally benign reaction paths

3. Source reduction via process integration

4. Establishing tradeoffs in pollution prevention

The ensuing sections provide a brief description of the literature available on these categories.

Design of separation systems for recycle and reuse of pollutants

A particularly attractive option for pollution prevention is the recycle and reuse of wastes via separation systems that can recover the pollutants. The task of designing a separation system for the recycle and reuse of pollutants can be stated as follows: Given a number of waste streams that contain certain undesirable species (pollutants), it is required to design a cost-effective network of separation units that can reduce the amount of the undesirable species in the waste streams to environmentally acceptable levels while recovering the pollutants in concentrations that are appropriate for recycle and reuse. For a given application, there are typically numerous options for undertaking the separation task. These alternatives are generated by considering the decisions involved in designing the system such as the selection of system components (e.g., which separation processes should be employed and which separating agents chosen?), system configuration (e.g., how should the units be arranged?), and operating variables (e.g., flow rates, temperature, pressure). In many cases there are numerous suboptions associated with each design decision, and an exhaustive enumeration of all the alternatives is prone to failure. Therefore, an efficient design strategy should attempt to identify the optimal solution while avoiding the enumeration of the vast number of alternatives. Mathematical programming techniques have been instrumental in this regard.

The incorporation of mathematical techniques into the design of recycle-and-reuse separation systems has attracted considerable research over the past few years. These research efforts have resulted in the development of systematic and generally applicable methodologies for the design of cost-effective separation systems that can recover the pollutants and deliver process streams in compliance with environmental regulations. In particular, three categories of separation processes have been considered: mass-exchange operations, heat-induced separations, and membrane-based systems.

The synthesis of *mass-exchange networks* (MENs) is probably the first attempt to systematically design recycle and reuse systems that employ *mass-separating agents* (MSAs) for the recovery of pollutants. An MSA is an additional species (e.g., solvent, adsorbent, ion-exchange resin) introduced into a process for the purpose of facilitating removal of contaminants. The MEN problem was introduced by El-Halwagi and Manousiouthakis[9–11] and can be stated as follows: Given a number of waste streams that are rich in some pollutants, synthesize a network of mass exchangers that can transfer the pollutants from the waste streams to the MSAs at minimum cost. A mass exchanger is any direct-contact mass-transfer unit that employs an MSA. Examples of mass-exchange systems include absorption, adsorption, solid-liquid extraction, liquid-liquid extraction, ion exchange, etc. The scope of the MEN problem has been generalized by Srinivas and El-Halwagi[12–14] to include reactive MSAs that can convert the pollutants to usable or salable species. Most MEN problems have been solved through targeting strategies. In particular, the mass-exchange pinch approach has provided powerful tools in tackling the problem.[9–14] The mass-exchange pinch concept is discussed in Chap. 8. The structural-optimization approach has also been successfully employed in solving the MEN problem.[15]

Chapter 16 illustrates the use of linear programming methods in the synthesis of MENs, using an example based on waste minimization in a styrene plant. A general wastewater minimization approach based on nonlinear programming methods is discussed in Chap. 17.

Heat-induced separation processes constitute another class of operations that can be employed in recycle and reuse. This group of processes includes condensation, crystallization, drying, and vaporization. In these systems, the pollutants are separated by phase change and can subsequently be recovered by mechanical separation, offering simple and clean operation. The problem of synthesizing *heat-induced separation networks* (HISENs) has been introduced by El-Halwagi et al.[16] and can be stated as follows: Given a number of waste streams that are rich in some pollutants, synthesize a cost-effective network of heat-induced separators that can reduce the concentration of the pollutants to environmentally acceptable levels via indirect heat transfer using energy-separating agents. This problem has been shown to be particularly useful for designing volatile organic compound (VOC) recovery networks from gaseous emissions. Mathematical procedures have been devised to solve the problem through a structural approach[17] as well as a decomposition (pinch-based) approach.[16–18]

Membrane-based operations are emerging separation technologies that have a significant potential for applicability in pollution prevention. Examples of these systems include reverse osmosis, pervaporation, and ultrafiltration. They are based on the principle of preferential permeability through membranes. In addition to their high selectivity (that can provide concentrations as low as parts per billion), low energy consumption and moderate cost, they are compact and modular. Therefore, membrane units can be readily added to existing plants to recover pollutants. The problem of designing a membrane separation network for recycle and reuse entails the identification of membrane types, sizes, number, and arrangement. In addition, the designer has to determine the optimal operating conditions and the type, number, and size of any pumps and energy recovery devices. Owing to the complexity of these systems, structural-representation approaches have been employed to solve this problem.[19,20.]

It is also worth pointing out that the recycle and reuse network may involve hybrid systems that include more than one type of separation technology. The synergism among various separation systems can provide major technical and economic benefits. Some research has been undertaken to integrate mass-exchange and membrane units[21] as well as mass-exchange and heat-induced operations.[22]

Synthesis of environmentally benign reaction paths

Owing to the substantial escalation in waste treatment costs, there has been a strong industrial trend to minimize waste generation at the source. To this end, a potentially attractive option is to alter the reaction chemistry in order to minimize the generation of undesirable by-products. The problem of synthesizing *environmentally acceptable reactions* (EARs) can be stated[23] as follows: Given a reactor of known size and functionality, and a number of desired products along with their flow rates, synthesize a set of chemical reactions that provide maximum gross profit while complying with all environmental and thermodynamic constraints. The problem complexity stems from the existence of numerous (almost infinite) reaction paths that can yield the desired products. Hence, a numerical approach is needed to address the problem effectively. The task of synthesizing EARs has been tackled by formulating the problem as an MINLP in which integer variables are used to account for the existence or absence of reactants and products.[23–25] These formulations have also benefited from the available techniques for reaction-path synthesis.[26]

Source reduction via process integration

Diminishing material and energy resources have spurred a strong industrial interest in process integration (see Chap. 4). One of the most developed techniques in process integration is thermal pinch analysis.[27] This technique is discussed in Part 2 of this book. Although pinch analysis was originally developed for heat recovery systems, it has recently been extended to assess the environmental impact of energy-related units. For instance, more effective heat integration can lead to lower duties of combustion units and, consequently, reduced gas emissions.[28–30] Numerical techniques of the type described in this chapter can enhance the usefulness of pinch analysis and other process integration methods. For example, optimization methods can be used to evaluate the economics of different pollution prevention options (including source reduction using pinch analysis) and hence to identify least-cost solutions to satisfy any given emission constraints.[31] This approach is discussed in Chap. 18.

Establishing tradeoffs in pollution prevention

The final category of numerical approaches to pollution prevention is based on multiobjective optimization. This class of problems is aimed at establishing tradeoffs among several objectives. For instance, one can characterize the relationship between the minimum cost for pollution prevention and the maximum extent of pollution control. Alternatively, the designer can determine the sensitivity of economic profitability to changing pollution prevention costs and regulations.[32]

The solution of multiobjective optimization problems typically entails the implementation of involved algorithms. Nonetheless, in some cases one can decompose the problem by solving it at several discrete levels, thereby allowing the use of simple computational tools such as spreadsheets.[31] Chapter 18 introduces a numerical graphical approach, combined with process integration concepts, for screening technical options for reducing gas emissions. The method can be readily implemented on a spreadsheet and can be used to identify least-cost solutions for any given level of process emissions. Another example on trading off source reduction against recycle and reuse is given in Chap. 16.

Real-Time Optimization of Process Operations

The focus of this book is the minimization of emissions through appropriate process design. However, the level of process emissions depends on not only design, but also the way a plant is operated. The same numerical methods that have been described in this chapter in the context of process design have been used successfully to optimize the use of equipment in existing facilities. Most often the objective of the optimization is to maximize profitability. However, it is possible to reformulate the problem and use the same methods to minimize emissions. For example, mixed-integer linear programming methods have been used to define ways of operating the steam power system of a chemical plant to minimize NO_x emissions in real time.[33] Numerical optimization can therefore play a role in minimizing emissions throughout the lifetime of an industrial process.

References

1. T. F. Edgar and D. M. Himmelblau, *Optimization of Chemical Processes*, McGraw-Hill, New York, 1988.
2. S. F. Hiller and G. J. Lieberman, *Introduction to Operations Research*, McGraw-Hill, New York, 1986.
3. G. V. Reklaitis, A. Ravindran, and K. M. Ragsdell, *Engineering Optimization*, Wiley, New York, 1983.
4. G. S. G. Beveridge and R. Schechter, *Optimization: Theory and Practice*, McGraw-Hill, New York, 1970.
5. R. Vaidyanathan R. and M. M. El-Halwagi, "Global Optimization of Nonconvex Nonlinear Programs via Interval Analysis," *Comp. ChE. Eng.*, 18(10):889–897, 1994.
6. V. Visweswaran and C. A. Floudas, "A Global Optimization Procedure for Certain Classes of Nonconvex NLP's—II. Application of Theory and Test Problems," *Comp. Chem. Eng.*, 14(2):1419–1434, 1990.
7. V. Manousiouthakis and D. Sourlas, "A Global Optimization Approach to Rationally Constrained Rational Programming," *Chem. Eng. Comm.*, 115:127–147, 1992.
8. N. V. Sahinidis and I. E. Grossmann, "Convergence Properties of Generalized Benders Decomposition," *Comp. Chem. Eng.*, 15(7):481–491, 1991.
9. M. M. El-Halwagi and V. Manousiouthakis, "Synthesis of Mass Exchange Networks," *AIChE J.*, 35(8):1233–1244, 1989.
10. M. M. El-Halwagi and V. Manousiouthakis, "Simultaneous Synthesis of Mass Exchange and Regeneration Networks," *AIChE J.*, 36(8):1209–1219, 1990.
11. M. M. El-Halwagi and V. Manousiouthakis, "Automatic Synthesis of Mass Exchange Networks with Single-Component Targets," *Chem. Eng. Sci.*, 45(9):2813–2831, 1990.
12. M. M. El-Halwagi and B. K. Srinivas, "Synthesis of Reactive Mass-Exchange Networks," *Chem. Eng. Sci.*, 47(8):2113–2119, 1992.
13. B. K. Srinivas and M. M. El-Halwagi, "Synthesis of Reactive Mass-Exchange Networks with General Nonlinear Equilibrium Functions," *AIChE J.*, 40(3):463–472, 1994.
14. B. K. Srinivas and M. M. El-Halwagi, "Synthesis of Combined Heat and Reactive Mass-Exchange Networks," *Chem. Eng. Sci.*, 49(13):2059–2074, 1994.
15. M. M. El-Halwagi and V. Manousiouthakis, "Simultaneous Design of Mass and Heat Exchange Networks," American Institute of Chemical Engineers annual meeting, Chicago, November 1990.
16. M. M. El-Halwagi, B. K. Srinivas, and R. F. Dunn, "Synthesis of Optimal Heat-Induced Separation Networks," *Chem. Eng. Sci.*, 50(1):81–97, 1995.

17. R. F. Dunn and M. M. El-Halwagi, "Selection of Optimal VOC-Condensation Systems," *J. Waste Management,* 14(2):103–119, 1994.
18. B. K. Srinivas, R. F. Dunn, and M. M. El-Halwagi, "Synthesis of Separation Networks with Simultaneous Heat Recovery and Waste Reduction," American Institute of Chemical Engineers annual meeting, St. Louis, November 1993.
19. M. M. El-Halwagi, "Synthesis of Optimal Reverse-Osmosis Networks for Waste Reduction," *AIChE J.,* 38(8):1185–1198, 1992.
20. B. K. Srinivas and M. M. El-Halwagi, "Optimal Design of Pervaporation Systems for Waste Reduction," *Comp. Chem. Eng.,* 17(10):957–970, 1993.
21. M. M. El-Halwagi, "Optimal Design of Membrane Hybrid Systems for Waste Reduction," *Sep. Sci. Tech.,* 28(1-3):283–307, 1993.
22. C. Stanley and M. M. El-Halwagi, "Practical Design Tools for Waste Management Systems via Process Integration," American Chemical Society 5th Symposium on Emerging Technologies in Hazardous Waste Management, Atlanta, September 1993.
23. E. W. Crabtree and M. M. El-Halwagi, "Synthesis of Environmentally-Acceptable Reaction Schemes," American Institute of Chemical Engineers annual meeting, St. Louis, November 1993.
24. J. P. Knight and G. McRae, "An Approach to Process Integration Based on the Choice of System Chemistry," American Institute of Chemical Engineers annual meeting, St. Louis, November 1993.
25. E. W. Crabtree and M. M. El-Halwagi, "Systematic Selection of Environmentally Benign Reaction Paths," American Chemical Society 5th Symposium on Emerging Technologies in Hazardous Waste Management, Atlanta, September 27–29, 1993.
26. E. Rotstein, D. Resasco, and G. Stephanopoulos, "Studies on the Synthesis of Chemical Reaction Paths," *Chem. Eng. Sci.,* 37(9):1337–1352, 1982.
27. B. Linnhoff and D. R. Vredeveld, "Pinch Technology Has Come of Age," *Chem. Eng. Prog.,* July 1984, pp. 33–40.
28. H. Korner, "Optimaler Energieeinsatz in der Chemischen Industrie," *Chem. Ing. Tech.,* 60:511–518, 1988.
29. H. D. Spriggs, R. Smith, and E. A. Petela, "Pinch Technology: Evaluate the Energy/Environmental Economic Trade-offs in Industrial Processes," Energy and the Environment in the 21st Century conference, Cambridge, MA, March 1990.
30. R. D. Sung, J. D. Kumana, and A. P. Rossiter, "Reducing Air Emissions by Process Heat Integration," American Society of Mechanical Engineers Ecoworld Conference, Washington, June 1992.
31. A. P. Rossiter, H. Klee, Jr., and R. D. Sung, "Process Integration Techniques for Pollution Prevention," American Chemical Society 4th Symposium on Emerging Technologies in Hazardous Waste Management, Atlanta, September 1992.
32. A. R. Ciric and S. Huchette, "Economic Sensitivity Analysis of Waste Treatment Costs in Source Reduction Projects: Discrete Optimization Problems," *Ind. Eng. Chem. Res.,* 32:2636–2646, 1993.
33. R. Nath and J. D. Kumana, "NO$_x$ Dispatching in Plant Utility Systems Using Existing Tools," 14th National Industrial Energy Technology Conference, Houston, April 1992.

16

Synthesis of Mass-Exchange Networks Using Linear Programming Techniques

Carol Stanley and Mahmoud M. El-Halwagi, Ph.D.
Department of Chemical Engineering
Auburn University
Auburn, Alabama

Introduction

As has been described in Chap. 4, the leading options in the pollution prevention hierarchy are source reduction and recycle and reuse. Source reduction is aimed at lowering the amount of pollutants generated at the source. On the other hand, recycle and reuse systems are employed to recover the polluting compounds for reuse in the plant or for sale as feedstock for other chemical processes. Hence, recycle and reuse systems are primarily composed of separation processes.

Over the past few years, there has been a consistent trend to progressively tighten environmental regulations. These regulatory changes require engineers to design more efficient recycle and reuse separation systems that remove pollutants down to minute concentrations. The procedure described in this chapter has successfully addressed these issues in pollution prevention case studies for several industries such as metal finishing,[1] inorganic fertilizers,[2] oil refining,[3] pulp and paper,[4] and synthetic fuels.[5] In many cases the procedure has identified more significant cost reductions than conventional engineering approaches have.

The greater the degree of separation required, the greater the cost involved and the more important it becomes to achieve a cost-effective design which meets environmental regulations. Additionally, the tight constraints on the degree of separation pose a variety of challenging questions. Should a single separation technology be used or a combination of two or more technologies? Out of a multitude of *mass-separating agents* (MSAs) that are available for a given separation, which one(s) should be used? What are the loads to be removed by each MSA? What is the flow rate of each MSA? Do any of the MSAs need to be regenerated? Which regenerating agents at what flow rates should be used? Is every stream leaving the network configuration environmentally acceptable, or is additional processing necessary? What is the network configuration?

The aforementioned questions indicate that the decisions involved in designing a recycle-and-reuse system are typically numerous. As has been mentioned in Chap. 15, there are virtually an infinite number of possible designs which must be screened economically. A trial-and-error procedure is prone to failure. The conventional heuristic methods of design are no longer adequate due to the unprecedented low concentrations which have not been encountered previously in typical industrial separations. Process synthesis, on the other hand, is ideally suited for this type of initial design work. Process synthesis allows the engineer to screen all possible configurations simultaneously and determine the configuration that minimizes the total cost of the system. Synthesis techniques can also be used to determine the cost or tradeoff involved with other configurations.

The remainder of this chapter will discuss the notion of synthesizing *mass-exchange networks* (MENs) as a useful framework for designing recycle-and-reuse systems. This framework can systematically provide answers to all the foregoing questions. In particular, it will be shown that the MEN synthesis approach can be coupled with linear programming techniques to establish the following:

- Optimum separation process(es)

- Flow rate of each MSA to minimize the operating cost of the network

- At least one system configuration that minimizes the fixed cost of the system while realizing the minimum operating cost

The synthesis of MENs will be described both in general terms and with an example from a styrene plant. Finally, a brief discussion will demonstrate the integration of source reduction and recycle and reuse.

Defining the MEN Synthesis Problem

The general MEN synthesis problem was first introduced by El-Halwagi and Manousiouthakis[6] and can be described as follows: Given a number of waste (rich) streams and a number of lean streams (MSAs), it is desired to synthesize a network of mass exchangers that can transfer certain species from the waste streams to the MSAs at minimum cost.

For each waste stream, the flow rate and supply (inlet) composition are known from the material balance on the process from which the waste streams originate. The target (outlet) composition of each waste stream is typically dictated by environmental regulations on emission standards for the pollutant in question. The supply (inlet) composition of each MSA corresponds to the concentration of the pollutant in the fresh or the regenerated MSA. On the other hand, the target compositions for the MSAs are determined based on a number of considerations which may be physical (e.g., solubility), technical (e.g., corrosion, fouling), environmental (e.g., regulations on emission standards), safety (e.g., flammability limits), and/or economic (e.g., to optimize the cost of subsequent regeneration of MSAs). In addition, the flow rate of each MSA is unknown and is to be determined so as to minimize the operating cost of the system.

A mass exchanger can be any direct-contact, countercurrent, mass-transfer operation that employs an MSA, such as absorption, adsorption, solvent extraction, leaching, and ion exchange. Potential MSAs include solvents, adsorbents, and ion-exchange resins. A schematic representation of the MEN problem is shown in Fig. 16.1.

Over the past few years, significant contributions have been made in solving general categories of the MEN synthesis problem. These categories involve the situations of MENs with a single pollutant,[1] multiple pollutants,[2] MSA regeneration,[7] reactive MSAs,[8,9] and nonisothermal operation.[10,11] The MENs have also been integrated with other pollution prevention systems such as membrane separations[12] and biological processes.[3]

The MEN synthesis procedure starts with a targeting phase in which the minimum cost of MSAs is established, without commitment to the network structure. This target is aimed at minimizing the operating cost of the system, since the cost of MSAs is typically the predominant operating expense in MENs. Hence, any design featuring the minimum cost of MSAs is referred to as a *minimum operating cost* (MOC) solution. Once the MOC solution is determined, a network is synthesized in which the number of mass exchangers is minimized while realizing the minimum cost of MSAs. Minimizing the number of mass exchangers provides an approximate method for minimizing the fixed cost of the network.

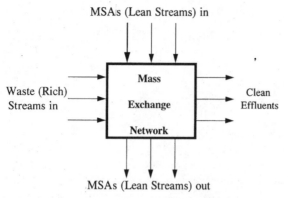

Figure 16.1   Schematic representation of MEN.

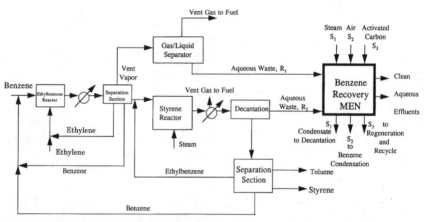

Figure 16.2   Schematic flow sheet of styrene plant.

Throughout the remainder of this chapter, the MEN synthesis procedure will be elucidated via a case study on recovering a *volatile organic compound* (VOC) from a styrene plant.

Case Study: Recovery of Benzene from the Aqueous Wastes of a Styrene Plant

A common process for producing styrene involves the dehydrogenation of ethylbenzene by using live steam over an oxide catalyst at temperatures of 600 to 650°C (see Fig. 16.2).[13,14] The first step in the process involves the production of ethylbenzene from ethylene and benzene. The reaction products are cooled and separated. Unreacted ethylene and benzene are recycled to the reactor. The vent vapor leaving the separation section is fed to a gas-liquid separator whereby an aqueous waste, R_1, is generated. The primary contaminant in this stream is benzene. Ethylbenzene leaving the separation section is forwarded to the styrene reactor. Styrene and hydrogen gas are the primary products formed in the reactor. Side products are benzene, ethane, toluene, and methane. The reactor product is then cooled in a condenser and sent to a decanter, where an organic layer and an aqueous layer form. The flue gases, consisting of hydrogen, methane, and ethane, do not condense in the condenser and are removed from the overhead of the condenser. The aqueous layer, consisting of steam condensate saturated with benzene, forms the second aqueous waste stream, R_2. The organic layer, consisting of styrene, benzene, toluene, and unreacted ethylbenzene, is sent to a series of distillation columns for separation. Recovered benzene is recycled to the ethylbenzene reactor, while recovered ethylbenzene is recycled to the styrene reactor.

There are two primary sources for aqueous pollution in this process: the condensate streams R_1 (1000 kg/h) and R_2 (69,500 kg/h). Both streams have the same supply composition, which corresponds to the solubility of benzene in water (approximately 1770 ppm or 1.77×10^{-3} kg benzene/kg water). Consequently, they may be combined as a single stream. The target composition is determined by the VOC environmental regulations.

In the last few years, there have been several major additions to the environmental regulations in the area of VOC emissions. These new regulations have greatly expanded the number of VOCs specifically regulated and have significantly lowered the allowable emission levels. The new VOC regulations in the United States apply to both air and wastewater emissions. For example, the benzene National Emission Standards for Hazardous Air Pollutants (NESHAPs) regulate the levels of benzene in both air emissions and wastewater. This regulation also requires control of "fugitive" benzene emissions from process equipment such as distillation columns, pumps, and storage tanks. The newest proposed regulations, which tackle the VOHAP (volatile hazardous air pollutant) and VVHAP (very volatile hazardous air pollutant), take the benzene NESHAPs a step further and apply to a large number of VOCs. Several VOCs have also been added to the *toxicity characteristic* (TC) rule, which is part of the hazardous waste regulations under the Resource Conservation and Recovery Act (RCRA). Chemical-specific effluent limitations are being added to wastewater standards under the National Pollutant Discharge Elimination System (NPDES). These are extremely stringent and apply to many VOCs. In the case of the styrene plant, the allowable emissions for benzene are 57 ppb in the wastewater effluent and 98 percent removal of benzene from any air emissions.[15,16]

The primary pollution prevention task in this case study is to synthesize a recycle-and-reuse system that can recover benzene from R_1 and R_2 and render an aqueous effluent with an environmentally acceptable benzene concentration of 57 ppb. In this example, three mass-exchange operations are considered: steam stripping, air stripping, and adsorption onto granular activated carbon, integrated as follows:

- The steam leaving the stripping unit is condensed and decanted. The condensate is recycled to the MEN while the benzene is recovered as an organic layer.

- The gaseous emission leaving the air stripper is fed to a cryogenic condensation unit, where 98 percent of the benzene is recovered and the air is discharged to the atmosphere.

- The spent activated carbon is regenerated by steam stripping. Hence, the separated benzene is recovered while the regenerated activated carbon is returned to the MEN.

The first step in screening the three proposed technologies for recovering benzene is to establish some criteria for the thermodynamic feasibility of waste recovery. These criteria are outlined next.

Establishing Thermodynamic Feasibility

One of the most essential steps in synthesizing an MEN is to ensure the thermodynamic feasibility of all mass-exchange duties. In the limiting case, the compositions of the pollutant in the waste stream and in each MSA would attain equilibrium. Since attaining equilibrium conditions requires the use of an infinitely large mass exchanger, it is important to maintain a minimum driving force for mass transfer. This driving force is referred to as the *minimum allowable composition difference* ϵ_j. This concept is demonstrated by Fig. 16.3. Any waste stream at a composition equal to or greater than Y can transfer the pollutant to an MSA whose composition is X or lower. Mathematically, the equilibrium relation governing the transfer of the pollutant from the waste stream to the jth MSA may be expressed as

$$Y^* = f_j(X_j^*) \tag{16.1}$$

where Y^* and X_j^* are the equilibrium compositions of the pollutant in the waste stream and the jth MSA, respectively, and f_j is the equilibrium function. When the minimum allowable composition difference ϵ_j is incorporated, this equation may be rewritten for a pair of compositions Y and X_j for which the thermodynamic and technical feasibility of mass exchange is guaranteed, as follows:

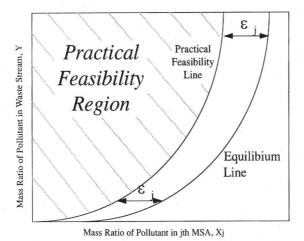

Figure 16.3 The practical feasibility region of mass exchange.

$$Y = f_j(X_j + \epsilon_j) \tag{16.2}$$

The practical feasibility line corresponds to Eq. (16.2). In addition, within the shaded region above the practical feasibility line, any operating line representing the material balance is thermodynamically and technically feasible.

It is worth discussing the significance of the minimum allowable composition difference ϵ_j. It represents the least acceptable mass-transfer driving force for the jth MSA (i.e., the smallest horizontal distance between the operating line and the equilibrium line on a McCabe-Thiele diagram). Hence, the smaller the ϵ_j, the greater the column height and, consequently, the higher the fixed cost. On the other hand, the smaller the ϵ_j, the lower the flow rate of the MSA and hence the lower the operating cost. In other words, the ϵ_j's are optimizable parameters that can be used to trade off capital against operating costs for the MEN. The designer can iteratively vary the ϵ_j's to attain the minimum total annualized cost of the network.[1]

The first step in synthesizing an MEN is to construct the *composition interval diagram* (CID). Figure 16.4 shows the CID for the styrene case study. In the CID a composition scale Y is established for the waste streams. Next, Eq. (16.2) is employed to generate a composition scale, X_j, for each MSA that is in one-to-one correspondence with Y. Hence, one obtains a total number of corresponding composition scales that is equal to the number of MSAs plus one. Then the waste streams and the MSAs are plotted as arrows on the diagram, using the appropriate composition scale for each stream. The head of the arrow corresponds to the

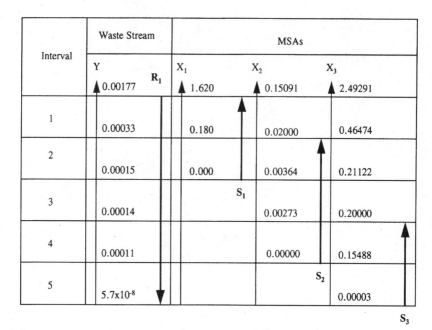

Interval	Waste Stream		MSAs		
	Y	R_1	X_1	X_2	X_3
	▲0.00177		▲1.620	▲0.15091	▲2.49291
1	0.00033		0.180	0.02000	0.46474
2	0.00015		0.000	0.00364	0.21122
3	0.00014		S_1	0.00273	0.20000
4	0.00011			0.00000	0.15488
5	5.7x10⁻⁸	▼		S_2	0.00003

S_3

Figure 16.4   Composition interval diagram for the styrene case study.

target composition of the stream and the tail of the arrow to the supply composition. Horizontal lines are drawn across the diagram for each head and tail of an arrow. These horizontal lines divide the diagram into composition intervals. Whenever a waste stream and an MSA are in the same interval, mass can be exchanged from the waste stream to the MSA. Also, mass can be exchanged from a waste stream in some interval to any MSA that lies below the interval of the waste stream.

The equilibrium data for benzene in the three MSAs and the cost data for the MSAs are given in App. A. The values of ϵ_1, ϵ_2, and ϵ_3 are taken as 0.15, 0.01, and 5×10^{-5} kg benzene/kg benzene-free MSA for steam stripping, air stripping, and activated carbon, respectively. The corresponding concentrations on the CID (Fig. 16.4) are derived from these equilibrium data and ϵ values.

The stream data for the MSAs are summarized in Table 16.1. As has been mentioned before, the supply (inlet) composition of each MSA corresponds to the concentration of the pollutant in the fresh or regenerated MSA. On the other hand, the target compositions for the MSAs are determined based on physical, technical, environmental, safety, and/or economic considerations. For instance, the supply composition of benzene in steam and air is zero since both streams enter the MEN as fresh MSAs. On the other hand, the supply composition of benzene in activated carbon corresponds to the concentration leaving the carbon regen-

TABLE 16.1   Stream Data for the MSAs

MSA	Supply composition (kg benzene/kg benzene-free MSA)	Target composition (kg benzene/kg benzene-free MSA)
Steam	0	1.620
Air	0	0.020
Activated carbon	3×10^{-5}	0.200

eration unit. As can be seen in Fig. 16.4, the target composition of benzene in steam is the maximum practically attainable composition that can operate against the inlet composition of the waste stream while maintaining a driving force of ϵ_1. The target composition of benzene in air is taken as 50 percent of the lower flammability limit. The target composition of benzene in activated carbon is set based on the saturation concentration of benzene onto activated carbon less the minimum allowable composition difference ϵ_3. Having established the stream data as well as thermodynamic feasibility criteria of mass exchange, we can now proceed to discuss the procedure for synthesizing the MEN.

Generating the MEN

As mentioned earlier, a targeting approach is used to synthesize the MEN. First, the minimum operating cost is identified by minimizing the operating cost of MSAs. Next, a network is developed by using the fewest mass-exchange units that can realize the MOC solution. The minimum allowable composition differences are used to trade off capital against operating cost.

One way of determining the MOC of an MEN is through the transshipment model, which is a classical optimization formulation. According to this model, one can conceptualize the MEN task as shipping a commodity (pollutant) from a number of sources (waste streams) to a number of destinations (MSAs) via intermediate warehouses (composition intervals).[1,8] Within any composition interval, it is feasible to transfer the pollutant from the waste streams to the MSAs. If the capacity of the MSAs in an interval is not sufficient to remove the load of the waste streams in the same interval, the residual mass can be passed down to a lower interval where it can be exchanged with MSAs at a more thermodynamically favorable composition.

A vanishing residual from an interval corresponds to the location of a thermodynamic bottleneck (mass-exchange pinch point). Any exchanger operating at the pinch point features the minimum allowable composition difference as a mass-transfer driving force at one end of the unit. The objective function of the model seeks to identify the flow rate of each MSA so as to minimize the operating cost of the network. The mathematical description of the optimization program is given in App. B. It is a linear program that can be solved globally by using commercial packages such as the ones mentioned in Chap. 15.

By employing this procedure, one obtains an MOC solution in which the optimal flow rates of noncondensing steam, air, and activated carbon are 70, 0, and 53 kg/h, respectively. The annual operating cost is about $15,000. Having determined the optimal flow rates of the MSAs, one can construct a network to realize the MOC solution. In many cases, it is possible to match streams based on their relative locations on the CID by noting that it is feasible to transfer mass horizontally or downward (i.e., to MSAs within the same composition interval or below). A general procedure for matching streams can be found elsewhere.[1,6] The network consists of a steam stripper, an adsorption column, and an activated-carbon regeneration unit. Steam stripping is used to remove about 90 percent of the benzene from the wastewater at a relatively high concentration of benzene. Subsequently, activated carbon is employed to carry out the remaining benzene recovery task at a relatively low concentration level of benzene. The MEN is schematically represented by Fig. 16.5. The *total annualized cost* (TAC) for the network is defined as

$$TAC = \text{operating cost} + f \times \text{capital cost}$$

By assigning a value of 0.2 for the annualization factor f one obtains the TAC for the network to be approximately $63,000/year. As has been

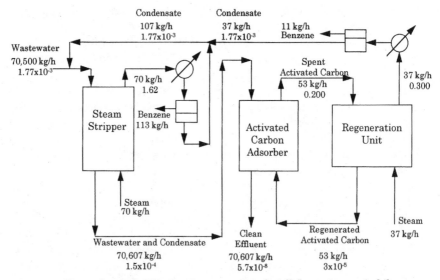

Figure 16.5 Network solution for the styrene case study (all flow rates are in kilograms of benzene-free stream per hour, numbers without units represent mass ratios, in kilograms of benzene per kilogram of benzene-free stream).

discussed earlier, the designer can trade off the operating cost against the capital investment by iterating over the ϵ_j's until the minimum TAC is achieved.[1] It is worth pointing out that the value of recovered benzene is about $219,000 per year, which significantly surpasses the TAC of the system. Hence, in addition to its positive environmental impact, waste recovery in this case provides an attractive revenue of $156,000 per year ($219,000 − $63,000).

Integration of Source Reduction with Recycle and Reuse

In many situations, there is a tradeoff between reducing the amount of generated waste at the source versus its recovery via a recycle-and-reuse system. For instance, consider again Fig. 16.2. Live steam is used in the styrene reactor to enhance the product yield. The higher the steam flow rate, the greater the ethylbenzene conversion. On the other hand, most of the steam eventually constitutes the aqueous waste R_2. Hence, the higher the flow rate of steam, the larger the cost of the benzene recycle and reuse MEN. These opposing effects call for the simultaneous consideration of source reduction of R_2 along with its recycle and reuse. One way of approaching this problem is by invoking economic criteria. Let us define the *economic potential* (EP) of the process as follows (all terms are expressed as dollars per year):

$$\begin{aligned}
\text{EP} = \ &\text{value of produced styrene} \\
&+ \text{value of recovered unreacted ethylbenzene} \\
&+ \text{value of recovered benzene} \\
&- \text{cost of ethylbenzene} \\
&- \text{cost of steam used in reactor} \\
&- \text{cost to separate unreacted ethylbenzene} \\
&- \text{TAC of recycle and reuse network}
\end{aligned}$$

Figure 16.6 demonstrates the dependence of the EP on the steam ratio (kg steam/kg ethylbenzene). The nonmonotonic behavior of this plot is indicative of the tradeoff between source reduction and recycle and reuse. As can be seen from Fig. 16.6, the optimum steam ratio is about 2.0.

Appendix A   Data for the Benzene Recovery Case Study

Equilibrium data

The equilibrium data for recovering benzene into steam, air, and activated carbon are given by

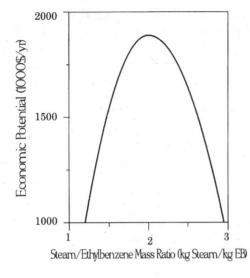

Figure 16.6 Economic potential of the styrene process versus steam/ethylbenzene ratio.

$$Y^* = m_j X_j^*$$

where Y^* and X_j^* are the equilibrium mass ratios (kg benzene/kg benzene-free stream) in the waste stream and the jth MSA, respectively. The index j takes the values of 1, 2, and 3 for steam, air, and activated carbon, respectively. The parameter m_j is the equilibrium distribution coefficient for benzene in the jth MSA. The values of m_1, m_2, and m_3 are 0.001, 0.011, and 7.1 × 10⁻⁴, respectively.

Cost data

The operating cost of steam stripping is taken as $4 × 10⁻³ per kilogram of noncondensing steam. This value is based on the price of low-pressure steam as well as the operating expenses involved in condensation and decantation. The operating cost of air stripping is $3 × 10⁻³ per kilogram of air. It includes the power needed to compress air to 2 atm and the cryogenic cooling used to condense 98 percent of the benzene. The operating cost of activated carbon is $0.026 per kilogram of recirculating activated carbon. It includes steam regeneration and makeup of activated carbon (assuming a continuous loss rate of 1 percent). It is worth pointing out that an additional cost is incurred in tackling the condensate resulting from steam stripping, cryogenic cooling of air, and activated-carbon regeneration. For steam stripping, the condensate leaving the decanter will contain a concentration of benzene equal to the solubility of benzene in water. Hence, the condensate stream should be returned to the MEN for recovery. When air stripping is used, it is necessary to dehumidify the benzene-laden air prior to cryogenic condensation to prevent icing. Typically, dehumidification is carried out by cooling the air to 274 K. This operation will result in a water condensate that contains the solubility limit of benzene in water. The flow rate of the condensate depends on the relative humidity of the airstream. For the styrene case study, we assume that each kilogram of air contains 0.016 kg of water vapor. Due to the benzene content of this condensate, it has to be forwarded to the MEN. In a similar manner, when activated carbon is regenerated, a steam/carbon weight ratio of 0.7 should be employed. When steam is condensed and decanted, it will yield an additional wastewater stream whose benzene concentration equals the solubility of benzene in water, and therefore this condensate stream should be recycled to the MEN. Since all three potential condensate streams, along with the wastewater leaving the styrene plant, have the same benzene composition, they should be mixed as a single stream entering the MEN. The additional expenses attributed to the cost of tackling the condensate streams will be accounted for in App. B.

The value of recovered benzene is taken as $0.20 per kilogram. The cost of ethylbenzene and the value of produced styrene are assumed to be $0.40 and $0.47 per kilogram, respectively.

The fixed cost of a moving-bed adsorption or regeneration column is given by $30,000V^{0.57}$, where V is the volume of the column (m^3) based on a 15-min residence time[17] for the combined flow rate of carbon and wastewater (or steam). The fixed cost of a sieve-tray stripper is $1750 per plate. The overall column efficiency is assumed to be 65 percent.

Appendix B   Mathematical Formulation

A linear programming formulation[1] is presented to identify the MOC solution of the MEN. As has been previously discussed, the first step in synthesizing an MEN is to construct the CID. Let us use the index k to designate the number of a composition interval, with $k = 1$ being the top interval and $k = N$ being the bottom interval. The kth composition interval is bounded by two horizontal lines. The set of corresponding composition scales at the upper line is denoted by $Y_{k-1}, X_{1,k-1}, X_{2,k-1}, \ldots,$ $X_{j,k-1}, \ldots, X_{NS,k-1}$, where NS is the number of MSAs. On the other hand, the set of corresponding composition scales at the lower line bounding the kth interval is referred to as $Y_k, X_{1,k}, X_{2,k}, \ldots, X_{j,k}, \ldots, X_{NS,k}$.

Within the kth interval, the load $W_{i,k}$ of the ith waste stream passing through this interval is determined from

$$W_{i,k} = G_i(Y_{k-1} - Y_k)$$

where G_i is the flow rate of the ith waste. Since G_i, Y_{k-1}, and Y_k are all known, the value of $W_{i,k}$ can be readily calculated. If waste stream i

does not pass through interval k, the value of $W_{i,k}$ is zero. Similarly, the capacity of the jth MSA in the kth interval is expressed as

$$w_{j,k} = L_j(X_{j,k-1} - X_{j,k})$$

when the jth MSA passes through the kth interval and as zero when it does not. The flow rate of the jth MSA, denoted by L_j, is unknown and to be determined via optimization.

The objective function for identifying the MOC solution can be expressed as

$$\min \sum_j c_j L_j$$

where c_j is the operating cost for the jth MSA (dollars per kilogram of the jth MSA). The set of constraints can be represented as follows:

- The material balance on the transferable pollutant around the kth composition interval is

Load of waste streams + surplus mass passed from interval $k - 1$ =
    capacity of MSAs + surplus mass passed to interval $k + 1$

  or

$$\sum_i W_{i,k} + \delta_{k-1} = \sum_j w_{j,k} + \delta_\kappa \qquad k = 1, 2, ..., N$$

- The surplus mass leaving each interval cannot be negative (second law of thermodynamics):

$$\delta_k \geq 0 \qquad k = 1, 2, ..., N - 1$$

$$\delta_0 = 0 \quad \text{and} \quad \delta_N = 0$$

- Flow rates of lean streams should be nonnegative:

$$L_j \geq 0 \qquad j = 1, 2, ..., NS$$

The following formulation for the styrene case study is included to illustrate this linear programming technique:

$$\min (0.004L_1 + 0.003L_2 + 0.026L_3)$$

subject to the following:

- Material balances around composition intervals:

$$14.4 \times 10^{-4} G_{\text{tot}} = \delta_1 + 1.440 L_1$$

$$1.80 \times 10^{-4} G_{\text{tot}} + \delta_1 = \delta_2 + 0.180 L_1 + 163.6 \times 10^{-4} L_2$$

$$0.10 \times 10^{-4} G_{\text{tot}} + \delta_2 = \delta_3 + 9.1 \times 10^{-4} L_2$$

$$0.30 \times 10^{-4} G_{\text{tot}} + \delta_3 = \delta_4 + 27.3 \times 10^{-4} L_2 + 451.2 \times 10^{-4} L_3$$

$$1.10 \times 10^{-4} G_{\text{tot}} + \delta_4 = 1548.5 \times 10^{-4} L_3$$

- Total flow rate of wastewater (equals flow rate from the process, 70,500 kg/h, plus any additional wastewater generated from steam stripping, air dehumidification, and activated-carbon regeneration):

$$G_{\text{tot}} = 70,500 + L_1 + 0.016 L_2 + 0.7 L_3$$

- Nonnegativity constraints:

$$\delta_1 \geq 0 \quad L_1 \geq 0$$

$$\delta_2 \geq 0 \quad L_2 \geq 0$$

$$\delta_3 \geq 0 \quad L_3 \geq 0$$

$$\delta_4 \geq 0$$

As has been mentioned earlier, the foregoing formulation is a linear program that can be readily solved by using commercial optimization packages such as the ones described in Chap. 15.

References

1. M. M. El-Halwagi and V. Manousiouthakis, "Automatic Synthesis of Mass Exchange Networks with Single-Component Targets," Chem. Eng. Sci., 45(9): 2813–2831, 1990.
2. M. M. El-Halwagi and V. Manousiouthakis, "Design and Analysis of Mass-Exchange Networks with Multicomponent Systems," American Institute of Chemical Engineers annual meeting, San Francisco, November 1989.
3. M. M. El-Halwagi, A. M. El-Halwagi, and V. Manousiouthakis, "Optimal Design of Dephenolization Networks for Petroleum-Refinery Wastes," Trans. Inst. Chem. Eng., 70B: 131–139, August 1992.
4. R. F. Dunn and M. M. El-Halwagi, "Optimal Recycle/Reuse Policies for Minimizing the Wastes of Pulp and Paper Plants," Env. Sci. Health, A28(1): 217–234, 1993.
5. B. K. Srinivas and M. M. El-Halwagi, "Optimal Design of Waste Management Systems for Synthetic Fuel Plants," American Chemical Society 4th Symposium on Emerging Technologies in Hazardous Waste Management, Atlanta, September 1991.

6. M. M. El-Halwagi and V. Manousiouthakis, "Synthesis of Mass Exchange Networks," *AIChE J.*, 35(8): 1233–1244, 1989.

7. M. M. El-Halwagi and V. Manousiouthakis, "Simultaneous Synthesis of Mass Exchange and Regeneration Networks," *AIChE J.*, 36(8): 1209–1219, 1990.

8. M. M. El-Halwagi and B. K. Srinivas, "Synthesis of Reactive Mass-Exchange Networks," *Chem. Eng. Sci.*, 47(8): 2113–2119, 1992.

9. B. K. Srinivas and M. M. El-Halwagi, "Synthesis of Reactive Mass-Exchange Networks with General Nonlinear Equilibrium Functions," *AIChE J.* 40(3): 463–472.

10. M. M. El-Halwagi, "A Process Synthesis Approach to the Dilemma of Simultaneous Heat Recovery, Waste Reduction and Cost Effectiveness," *Proceedings of the Third Cairo International Conference on Renewable Energy Sources*, A. I. El-Sharkawy and R. H. Kummler, eds., Academy of Scientific Research and Technology, Cairo, vol. 2, 1993, pp. 579–594.

11. M. M. El-Halwagi and V. Manousiouthakis, "Simultaneous Synthesis of Heat and Mass-Exchange Networks," American Institute of Chemical Engineers annual meeting, Chicago, November 1990.

12. M. El-Halwagi, "Optimal Design of Membrane Hybrid Systems for Waste Reduction," *Sep. Sci. Tech.*, 28(1-3): 283–307, 1993.

13. C. A. Brighton, G. Pritchard, and G. A. Skinner, *Styrene Polymers: Technology and Environmental Aspects*, Applied Science Publishers, London, 1979, pp. 15–28.

14. P. J. Lewis, C. Hagopian, and P. Koch, "Styrene," in *Kirk-Othmer Encyclopedia of Chemical Technology*, 3d ed., vol. 21, M. Grayson, ed., Wiley, New York, 1983, pp. 770–801.

15. U.S. Environmental Protection Agency, 40*CFR*, Section 414, Subparts F and I, 1992, pp. 276–278 and 285–286.

16. U.S. Environmental Protection Agency, 40*CFR*, Section 61, Subpart FF, pp. 162–189, as amended in *Fed. Reg.*, 58(4):3072–3105, 1993.

17. M. H. Stenzel and W. J. Merz, "Use of Carbon Adsorption Processes in Groundwater Treatment," *Environ. Prog.*, 8(4): 257–264, 1989.

Wastewater Minimization Using Nonlinear Programming

Alan P. Rossiter, Ph.D.
Linnhoff March, Inc.
Houston, Texas

Ravi Nath, Ph.D.
Setpoint, Inc.
Houston, Texas

Introduction

In recent years there has been a growing trend toward *zero discharge* for wastewater from processes.[1] In some cases this has even become a legal requirement. In South Africa, e.g., a zero discharge regulation prohibits certain industrial complexes from disposing of water that has been in contact with their processes, products, or raw materials into natural waterways.[2]

Process integration has a significant part to play in minimizing wastewater quantities and achieving or approaching zero discharge. The *mass-transfer pinch* is one of the applicable approaches, and this is described in Chap. 8 of this book. This chapter discusses another approach—the use of *nonlinear programming* (NLP) methods to identify *minimum wastewater* designs for processing systems.

To use this approach, a component mass balance model is set up. This defines a superstructure incorporating all relevant water streams within the process under investigation, together with all operations that introduce or remove waterborne contaminants. A means of characterizing wastewater streams is introduced, and this provides a convenient basis for defining appropriate options for recycling or reusing each stream. This information is then processed by a commercial optimization package. The output from the procedure comprises two main elements:

- Target minimum wastewater flow

- One possible process structure for achieving the target flow

This approach has been used successfully in studies of facilities in a number of industrial sectors, including fibers, pulp and paper, and oil refining. Typical reductions in freshwater use and wastewater discharges identified are around 30 percent. The procedure is illustrated in this chapter by reference to an oil refining application.

Defining the Problem

Wastewater sources in industrial processes fall into three major categories:

1. Water is introduced into a process deliberately to remove impurities that would otherwise disrupt processing operations (e.g., desalting in oil refining, brownstock washing in Kraft pulp mills, and various solvent extraction operations). This creates a wastewater stream that contains the impurities and removes them from the process.

2. The wastewater is an unavoidable by-product of process operations (e.g., decanted water removed from crude oil or water generated by reaction processes).

3. The wastewater is a by-product of utility water use (e.g., blowdown from boilers or cooling water systems).

The first type of wastewater generation (i.e., deliberate water addition to remove impurities) is typically characterized by a requirement to remove a *fixed amount* of one or more contaminants (e.g., salt in a crude oil desalter). It is subject to maximum contaminant concentrations for water entering and leaving the unit in which the mass transfer occurs. However, in certain cases, there is *no fixed quantity* of one or more contaminants to be removed. In the desalter, e.g., the wastewater (brine) contains some oil—an undesirable but inevitable consequence of contacting oil and water. It is the *bulk concentration* of the oil in the wastewater, rather than the absolute quantity, that is typically constant for steady desalter operation.

The second type of wastewater generation (generated by process operations in which water is *not* added specifically to remove contaminants) typically results in a fixed mass flow of *secondary water,* with fixed contaminant concentrations. Neither the flow nor the composition is generally amenable to modification. However, in some cases, the secondary water can be used in place of fresh water in operations that use water.

The third type of wastewater generation (blowdown from utilities) is generally associated with the concentration of trace impurities. This arises because makeup water (which always contains some impurities) has to be added to the utility system to replace lost water, e.g., evaporative losses from cooling towers, steam losses (due to leaks and live steam injection), and condensate losses from steam systems. The losses are generally substantially pure water, so in the absence of a blowdown the contaminants from the makeup would concentrate to unacceptable levels within the utility system. Chemicals (e.g., biocides) are also commonly added to water within utility systems, and the blowdown stream inevitably contains some of these materials.

Solving the Problem

The procedure focuses on reuse, regeneration, and recycle of wastewater. These terms are defined here as follows:

- *Reuse:* taking contaminated water from one operation and using it as feed to another

- *Regeneration:* treating contaminated water to make it suitable for use in one or more of the water-consuming operations

- *Recycle:* returning water (via a regenerator) to the unit that produced it

Regeneration

There are many different types of regeneration equipment, such as filters, membrane separators, sour water strippers, and ion-exchange systems. The objective of all these is to remove contaminants to make the water stream suitable for reuse or recycle.

In almost all regeneration operations (e.g., membrane systems) the incoming contaminated water is split into two streams, one "pure" (i.e., suitable for reuse or recycle) and the other containing the contaminants at increased concentration. Thus only part of the incoming flow can be reused. Moreover, most regenerators have to be "regenerated" themselves; e.g., filters require periodic backflushing, and ion-exchange resins require restoration to their original ionic content. This regeneration of the regenerators typically requires additional use of fresh water and thus increases both the use of fresh water and the total wastewater flow.

Model and solution method

The model is summarized in the appendix. This represents mass transfer in dilute aqueous solutions within each unit operation. The user must specify all units that generate wastewater (including secondary water sources) and all regeneration options. The model automatically identifies all possible recycle and reuse options for each water stream.

In its simplest form, the wastewater minimization problem is bilinear. It is not readily amenable to linear approximations. For this reason, nonlinear optimization methods are needed. Commercial NLP packages have been found to solve the model satisfactorily and to identify which of the possible options minimizes the overall wastewater flow subject to any additional user-specified constraints.*

The mathematics of nonlinear optimization does not guarantee a *global* optimum. The solution that is found may be dependent on the initialization and is in general a *local* optimum, which may or may not also be the global optimum. In practice, it has been found that a global optimum is easy to locate for wastewater problems in which there is no regeneration. Where regeneration is included, the problem is more difficult, and multiple initializations may be necessary. However, "realistic" initializations, i.e., initializations that correspond to reasonable, practical designs, generally lead to the true global optimum. Appropriate initializations may be developed from a knowledge of the physics and chemistry of the process or by using insights obtained from other process integration methods [notably hierarchical review (Chap. 12) or mass-transfer pinch analysis (Chap. 8)].

The NLP algorithm seeks to identify the solution that minimizes the objective function—in this case, wastewater flow (or fresh makeup water). If the global optimum has been reached, the value of the objective function at the solution is the minimum feasible water flow. However, the structure identified by the optimizer is not necessarily unique; i.e., other structures may be possible that achieve the same target flow.

Another useful output that can be obtained from most commercial optimization packages is *shadow pricing*. This provides the partial derivatives of the objective function at all active constraints. These can be very helpful. In practice, many constraints (e.g., allowable contami-

---

*The authors used the GINO NLP package, from LINDO Systems, Inc., P.O. Box 148231, Chicago, IL 60622. The model and user interface are coded in a package called WATERLILY, developed by Linnhoff March, Inc., 9800 Richmond Ave., Houston, TX 77042.

nant concentrations in certain pieces of equipment) are not known with great accuracy. Shadow pricing data tell the engineer where the most significant constraints are and what the incentives are if those constraints can be relaxed. This information can thus be a useful guide for research efforts.

Case Study

The wastewater minimization technique is illustrated by an oil refinery example. Both the (simplified) structure of the wastewater system and the stream flows (Fig. 17.1) are loosely based on part of Amoco's Yorktown refinery[3] (see also Chaps. 12 and 21). Concentrations assumed for contaminants in the different streams are based on data from various literature sources.[4,5]

The wastewater system handles effluent from the *desalter*, together with *sour wash water*. There are also additional miscellaneous *sour water* streams. For convenience, these are combined in the present analysis, and they form a secondary water stream. A *sour water stripper* (SWS) is used to regenerate the sour water. There is also a *cooling water system*. The italicized terms are discussed below. Data for this system are summarized in Table 17.1.

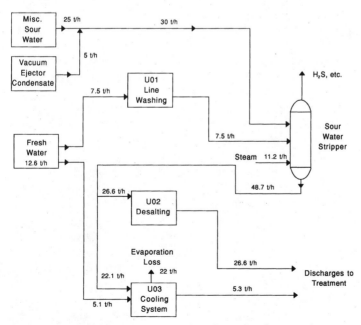

Figure 17.1   Refinery case study: initial configuration (base case).

TABLE 17.1   Data Set for Refinery Wastewater Example

Process	Water loss (t/h)*	Contam- inants	Max. C_{in} (ppm)	Max. C_{out} (ppm)†	Pickup (kg/h)
Line washing		Hydrocarbon	50	120^E	n/a
		H_2S	5,000	$12,500^E$	n/a
		Salt	140	200	1.5
Desalter		Hydrocarbon	120	220^E	n/a
		H_2S	20	45^E	n/a
		Salt	200	9,500	250
Cooling system	22.0	Hydrocarbon	220	0^E	n/a
		H_2S	45	0^E	n/a
		Salt	2,500	3,115	13.86‡

*1 t/h = 0.278 kg/s.
†The E implies that the outlet concentration is a fixed (equilibrium) value.
‡Pickup represents solids in fresh makeup water. The software used in this study assumes zero contamination in the fresh makeup water. In reality, there would be 630 ppm of hardness in this water. The pickup of 13.86 kg/h of solids in the cooling system model represents the quantity of hardness introduced if fresh water is used to replace the 22.0 t/h evaporative loss.
   Secondary water. The secondary water (sour water) flow is 30 t/h. It contains 45, 4000, and 135 ppm of hydrocarbon, H_2S, and salt, respectively. Excess secondary water (i.e., any that is not used within the unit operations) will require end-of-pipe treatment.
   Sour water stripper. The sour water stripper is assumed to remove 0, 99.9, and 0 percent of the hydrocarbon, H_2S, and salt, respectively, with no water in the overheads.

Desalter and sour wash water

Desalting is carried out to remove unwanted dissolved and suspended materials from the crude. This process involves contacting water with the incoming oil and then decanting it. Wash water is used to prevent the buildup of salts in vapor lines around the refinery, notably in hydrotreating units. Because it is typically in contact with sour gases, the wash water effluent is generally also sour (see discussion on sour water below).
   The primary purpose of both the desalter and washing is the removal of definable and relatively small quantities of specific contaminants (mostly salts). However, additional contaminants (notably hydrocarbons and H_2S) are picked up because the desalting water and wash water have to be contacted with liquid and/or vapor streams that contain large quantities of hydrocarbons and H_2S. Small amounts of these contaminants are consequently either dissolved or suspended in the water. This results in a roughly constant concentration of these contaminants in the effluent water, irrespective of water flow. The data in Table 17.1 reflect the ways in which the different contaminants behave: Salt contamination is defined in terms of maximum permissible inlet and outlet concentrations in the water stream and a fixed amount of pickup. For the two other contaminants—H_2S and hydrocarbons—maximum permissible inlet concentrations are specified, but the outlet con-

centration is fixed by equilibrium considerations. There is *no pickup specification,* because the amount of these contaminants transferred to the water will depend on the inlet concentration and flow rate.

Sour water

Sour water is any water contaminated with H_2S. Its composition may vary considerably, depending on its source. In the present simplified example, it is assumed that all the sour water sources (excluding the sour wash water) are combined into a single "miscellaneous sour water" stream of fixed composition. Sour water is produced in a number of ways, e.g.:

- *Steam ejectors.* Steam ejectors are used to provide the vacuum for the vacuum distillation unit. The condensate obtained from these systems contains hydrocarbons and H_2S.

- *Column condensers.* Water in column feeds and live steam used for stripping tend to go overhead in distillation columns and have to be drawn off. These condensates are often contaminated with H_2S.

- *Storage tank drains.* Water also has to be drawn off tanks in which feed and product materials are held. Depending on the nature of the material being stored, this may be contaminated with varying amounts of H_2S, hydrocarbons, and dissolved and suspended solids.

Sour water stripper (SWS)

Sour water stripping is the means most commonly adopted for removing H_2S from refinery sour waters. This is necessary both to recover the H_2S for removal of elemental sulfur and to make the water suitable for reuse or disposal. It is accomplished in a stripping column, by using either live steam injection or reboiling. In the present example, the existing plant has a live steam system. The live steam that is injected condenses and mixes with the stripped water. Consequently the water flow leaving the stripper is roughly 30 percent more than the water flow entering. The effluent water is used in the desalter, with excess water going to the cooling system.

Cooling water system

The cooling system differs from the other water users in that it has a water loss of 22.0 t/h associated with it, due to evaporative cooling in the cooling tower. Cooling towers act as strippers for light components in water. For simplicity in this example, this stripping action is modeled as an equilibrium system for the hydrocarbons and H_2S, with zero outlet concentration for both species.

All cooling tower makeup water contains dissolved and/or suspended solids—even if the makeup water is notionally "pure." Evaporation from the cooling tower tends to concentrate these solids, and it is necessary to blow down to prevent buildup that would otherwise lead to scaling. Calcium and magnesium compounds, in particular, contribute to this problem. A typical value for the "hardness" of fresh makeup water is around 630 ppm. Blowdown concentrations are typically around 5 times this value.[5] (In the present example, no distinction is made between different solids species; neither is any distinction made between dissolved and suspended solids. For a more rigorous evaluation, additional contaminant species can be incorporated in the data set to account for these.)

Water treatment chemicals, notably biocides, are added to cooling water systems. These generally make the blowdown unsuitable for reuse.

Base-case water uses

Figure 17.1 represents the *base case* for this study, in terms of both structure (i.e., equipment items and their interconnections) and water flows. The demand for fresh water is 12.6 metric tons per hour (t/h), split between line washing applications and the cooling system. All the "miscellaneous sour water" (5 t/h derived from steam used in vacuum ejectors and 25 t/h from other sources; 1 t/h = 0.278 kg/s or 2204.6 lb/h) is fed to the SWS, together with line washing effluent (7.5 t/h) for a total regeneration feed rate of 37.5 t/h. Of live steam 11.2 t/h is used in the SWS, and this condenses and becomes a part of the SWS effluent stream. The principal use for this stripped sour water is the desalter (26.6 t/h), with the excess (22.1 t/h) going to the cooling tower.

In practice, the excess stripped sour water is problematic, because it contains small amounts of phenols. These create an odor in the vicinity of the cooling tower. It would not be acceptable to discharge this stream outside the refinery, and there is no other potential user for this water within the refinery. A reduction in the volume of the excess would therefore be highly desirable.

In terms of overall water balance, the live steam (to the SWS and the vacuum ejectors) is additive with the fresh water, so the total *clean water* supplied (as steam or fresh water) is 12.6 + 11.2 + 5.0 = 28.8 t/h.

Flow optimization without process changes

The data set for this process is summarized in Table 17.1. It is assumed that all sour water streams must go directly to the sour water stripper, and this is accepted as a constraint.

Figure 17.2 shows the structure identified by the optimizer. The main structural difference relative to the base case (Fig. 17.1) is that

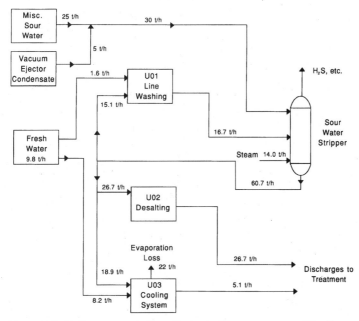

Figure 17.2 Refinery case study: optimized with no process modifications.

stripped sour water provides most of the flow (15.1 t/h) for line washing, with only a small amount (1.6 t/h) of fresh water. The fresh water is needed because the inlet hydrocarbon concentration constraint (50 ppm) would be exceeded without it.

The *clean water* flow (fresh water plus steam) is 28.8 t/h. This is the same as the use in the existing design. However, the stripped sour water to the cooling tower goes down from 22.1 to 18.9 t/h, a reduction of 3.2 t/h, or 14 percent, which would slightly alleviate the cooling tower odor problem. There is a cost for this, however: The regeneration feed rate goes up to 46.7 t/h (from 37.5 t/h), and live steam use in the SWS rises to 14.0 t/h (from 11.2 t/h). These changes represent 25 percent increases. There is a reduction in freshwater use (offsetting the increased live steam demand). However, steam is far more expensive than fresh makeup water (typically around $8.00 per metric ton versus $0.50 per metric ton), so this change results in a net operating cost *increase* of around $168,000 per year.

Flow optimization with process changes

The study of the Amoco Yorktown refinery using the hierarchical review procedure (see Chap. 12) came up with a list of several potential means of reducing emissions, including a number of ideas that would reduce stripped sour water flow to the cooling tower. Two of these were to

1. Convert the SWS from live steam to a reboiled system (to reduce the aqueous effluent flow from the SWS).
2. Replace vacuum steam ejectors with vacuum pumps (to reduce the quantity of sour condensate produced).

These ideas can be evaluated by using the optimization procedure. The results are shown in Fig. 17.3 (SWS modification) and Fig. 17.4 (SWS and ejector modifications).

Replacing the live steam flow in the SWS with a reboiler (Fig. 17.3) means that there is far less excess stripped sour water to send to the cooling system (8.3 t/h, down from 18.9 t/h). This should result in another significant improvement in the odor problem. The clean water target is 23.7 + 5.0 = 28.7 t/h, slightly down from the situation in Fig. 17.2 (28.8 t/h, including water and steam). However, in Fig. 17.3 all the demand (except for the vacuum ejectors) is in the form of fresh water rather than steam. (There is still a need for steam to drive the SWS, but the condensate does not mix with the sour water. Provided the conden-

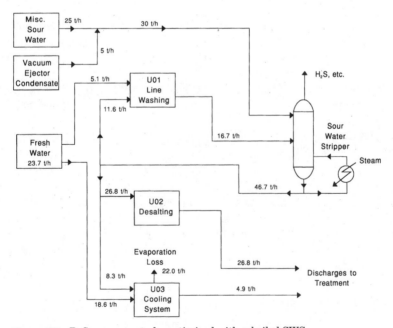

Figure 17.3 Refinery case study: optimized with reboiled SWS.

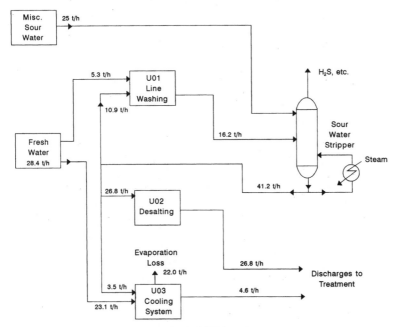

Figure 17.4 Refinery case study: reboiled SWS and no steam ejectors.

sate is recovered and returned to boilers, the steam does not contribute
to freshwater demand.) The flow entering the SWS is still 46.7 t/h, as
in Fig. 17.2.

The ejectors contribute 5 t/h to the sour water flow. If they are
replaced with vacuum pumps, the total miscellaneous sour water flow
goes down from 30 to 25 t/h. There is a corresponding reduction in the
demand on the SWS, which will reduce steam costs. If this change is
made together with the introduction of a reboiler on the SWS (Fig.
17.4), then the flow to the SWS goes down to 41.2 t/h and the stripped
sour water flow to the cooling system goes down to 3.5 t/h, further
reducing the odor problem. However, the *freshwater* demand increases
(from 23.7 to 28.4 t/h). This is simply a water balance issue: Eliminating 5 t/h of sour water derived from steam condensate in the vacuum
ejectors creates a need for (approximately) 5 t/h of water from another
source. Of course, there has also been a reduction of approximately 5
t/h in steam for steam ejection.

One final process modification is now considered: Up to this point it
has been accepted that the *miscellaneous sour water* stream has to go
to the sour water stripper. If this restriction is removed, the result
shown in Fig. 17.5 is obtained. The bulk of the miscellaneous sour

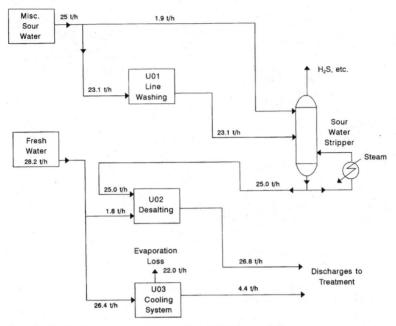

Figure 17.5 Refinery case study: reboiled SWS and no steam ejectors; miscellaneous sour water to line washing.

water now goes to line washing. There is a very small reduction in freshwater demand (28.4 to 28.2 t/h). More importantly, the flow to the sour water stripper decreases from 41.2 to 25 t/h, which greatly reduces steam costs; and the stripped sour water excess flow to the cooling system is totally eliminated. A small amount of fresh water is now required for the desalter.

Results

The results of the case study are summarized in Table 17.2. In this particular case, there is very little reduction in the overall wastewater flow from the plant (approximately 2 percent), because the stripped sour water has to be replaced by fresh makeup water in the cooling tower. However, there is a 100 percent reduction in the excess stripped sour water flow (which eliminates the cooling tower odor problem) as well as a 33 percent reduction in the sour water stripper feed flow (which reduces the operating costs of the sour water stripper by approximately $220,000 per year). There is also a small operating saving through replacement of the vacuum ejector with a vacuum pump. It is difficult to put a monetary value on the environmental improvement in the vicinity of the cooling tower.

TABLE 17.2 Refinery Case Study Results

| | Water flows (t/h)* | | | |
Case description	Live steam (SWS and vacuum ejector)	Fresh water	Stripped sour water to cooling tower	SWS feed
Base case	16.2	12.6	22.1	37.5
Optimized				
No process changes	19.0	9.8	18.9	46.7
Reboiled SWS	5.0	25.7	8.3	46.7
Reboiled SWS and no steam ejectors	0.0	28.4	3.5	41.2
Reboiled SWS, no steam ejectors, SW to line wash	0.0	28.2	0.0	25.0
Additional benefits:				
Eliminate cooling tower odor				
Reduce operating costs by $220,000 per year				

*1 t/h = 0.278 kg/s.

Comparison with the Concentration Pinch Method

Mass-transfer pinch analysis (Chap. 8) and the NLP procedure described in this chapter are both methods for minimizing wastewater flows. They are based on very different solution techniques and are useful for rather different reasons. However, if the same data and constraints are used, the results obtained by the two methods should be identical. In practice, differences are sometimes found. This can be the result of either failure of the NLP algorithm to locate the global optimum or the limitations of the pinch calculations for dealing with multicontaminant problems. Discrepancies can also arise because different constraints are inadvertently applied when the two methods are used. The two approaches are compared and contrasted below.

Main purpose

The mass-transfer pinch approach is primarily a conceptual tool, with a largely graphical output. It provides insights into design options that are not readily obtained in any other way. The NLP approach, in contrast, is basically a number-crunching method. It yields a rigorous solution to an optimization problem, but gives few insights into design options.

Targeting and design tasks

The minimum water flow generated by the NLP procedure corresponds to the *target* flow in mass-transfer pinch analysis, and the structure provided in the output corresponds to the *design* obtained in mass-transfer pinch analysis. The NLP procedure therefore generates both the target and a feasible design to achieve the target in a single step. In mass-transfer pinch analysis, targeting and design are distinct tasks.

Guarantee of optimized solution

The mass-transfer pinch approach results in a unique water flow target for any given set of data, and this represents a guaranteed mathematical minimum (subject to the assumptions built into the analysis). Nonlinear programming, in contrast, guarantees only a local optimum. Therefore there is always the possibility that the solution obtained by this method is not the true *global* minimum for water flow. In practice, provided there is a good initialization for the model, the solution actually obtained is generally either at the global optimum or very close to it.

Handling of multicomponent problems

With the NLP approach, it is not significantly more difficult to deal with problems involving arbitrarily large numbers of components than it is to deal with single-component problems, provided the optimization package and computer are sufficiently powerful. With mass-transfer pinch analysis, however, multicomponent analysis requires mathematical transformations, which make it rather difficult (especially for a nonexpert user) to interpret the results.

Constraints

Numerical programming methods are designed specifically to deal with constraints. In the context of wastewater minimization, this makes it very easy to forbid or force connections between operations (e.g., to force the miscellaneous secondary water to go to the sour water stripper—but nowhere else—in the example in this chapter). This is much more difficult to do with the mass-transfer pinch approach.

Composition rules

With the NLP approach it is very easy to write virtually any equation to define concentration relationships [e.g., the fixed pickup and fixed bulk concentration (equilibrium) models described earlier in this chapter]. In principle, the pinch-based approach can also handle almost any pickup relationship. In practice, however, the fixed pickup model is by far the easiest to implement.

Regeneration recycle and regeneration reuse

One useful feature of the concentration pinch method is that it can generate water consumption targets for regeneration reuse and regeneration recycle.[4] In the reuse case, regenerated water is not permitted to reenter process operations in which it has previously been used, while in the recycle case it can reenter these operations. The reuse targets can be useful in cases where built-up trace contaminants that are not removed in regeneration would be problematic.

It is difficult to prevent recycle without introducing other (unintended) constraints in the NLP approach, especially in large problems. However, it is relatively easy to model concentration buildup in recycle systems, so the NLP approach can deal with most practical problems where the buildup of impurities is an issue.

Targets

There is a fundamental difference in the way in which the data are handled in the two methods, and this can result in significant differences in targets. The NLP approach treats all operations as discrete, indivisible entities, in which only the inlet and outlet conditions are meaningful. The mass-transfer pinch approach, in contrast, treats operations as having continuous variations in composition between their inlets and outlets.

The *indivisibility* of operations in the NLP approach is a constraint. Consequently, if the water flow can be reduced by splitting an operation, the pinch target is lower than that generated by the NLP approach. In most practical situations, however, the two targets are very similar.

Designs

The difference in the way in which the data are handled results in rather different design methods. With mass-transfer pinch analysis, there are two approaches possible in design. The first makes use of mass-transfer driving forces, and this can often lead to the requirement that an operation be split. The second approach guarantees that a single source of water is used for each operation—in other words, operations do not need to be split. However, when regeneration is introduced, the splitting of operations may be necessary to achieve the target, even with the second method. The design can always be modified to remove the split by increasing the water flow by an appropriate amount.[6]

In the programming approach, because each operation is treated as discrete and indivisible, the structures that are generated always perform each operation in a single unit. Consequently, little (if any) structure evolution is needed to simplify the design.

Summary and Conclusions

Nonlinear programming provides a powerful means of quantifying the scope for wastewater minimization and identifying specific recycle and reuse arrangements to achieve the target. The method provides a rigorous solution for the minimum freshwater demand or wastewater generation rate in a production facility, and it can readily be incorporated in a user-friendly software package. The technique is particularly well suited to situations where some prescreening of process options has already been carried out, to identify appropriate regeneration options and process modification opportunities. Prescreening can be accomplished with one of the other process integration methodologies, such as hierarchical review or mass-transfer pinch analysis. Industrial studies using this approach have typically identified opportunities to reduce water flow by around 30 percent.

Acknowledgments

The authors wish to acknowledge the contributions of Jimmy Kumana of Linnhoff March, Houston, Texas, and Ammi Amarnath of the Electric Power Research Institute (EPRI), Palo Alto, California, in helping to develop the approach described in this chapter.

Appendix   Water Minimization Model

The model described here is set up to handle up to 20 unit operations, six contaminants, three regenerators, and four water supplies. One of the water supplies is required to be fresh water with zero contaminant concentration.

The model includes two alternative contaminant pickup modes (equilibrium or fixed quantity) and two different regeneration modes (fixed percentage removal and fixed outlet concentration). Alternative objective functions (minimize fresh water or minimize wastewater) are also provided. A commercial optimization package can be used to solve the model and identify the conditions that minimize the objective function.

Variables

$S_{s,i}$ = secondary water flow to unit operation i, from source s

W_i = primary water flow to unit operation i

$B_{i,k}$ = unit operation i ($i = 1,...,20$) inlet water, contaminant k concentration

$C_{i,k}$ = unit operation i ($i = 1,...,20$) outlet water, contaminant k concentration

$C_{r,k}$ = regenerator r ($r = 21,...,23$) exit stream contaminant k concentration

$X_{i,j}$ = water reuse from unit i ($i = 1,...,20$) to unit j

$X_{i,r}$ = water flow from unit i ($i = 1,...,20$) to regeneration r ($r = 21,...,23$)

$X_{r,i}$ = water flow regeneration r ($r = 21,...,23$) to unit operation i ($i = 1,...,20$)

$X_{r,T}$ = water flow regeneration r ($r = 21,...,23$) to treatment

L_i = water loss (evaporation) from unit i

Constants

$\text{ctype}_{i,k}$ = contaminant type (E = equilibrium: outlet concentration fixed; P = pickup fixed)

$\text{bmax}_{i,k}$ = maximum inlet concentration of contaminant k to unit operation i

$\text{cmax}_{i,k}$ = maximum outlet concentration of contaminant k from unit operation i

$\text{pickup}_{i,k}$ = contaminant k pickup in unit operation i

$k1_r$ = multiplier for water return to process from regenerator r

$k2_r$ = multiplier for water return to treatment from regenerator r

$\text{cs}_{s,k}$ = concentration of secondary water, component k

$\text{rmode}_{r,k}$ = regeneration model for component k (fixed percentage removal or fixed outlet concentration)

$\text{rspec}_{r,k}$ = regeneration model specification (percentage removal or outlet concentration)

smax_s = maximum amount of secondary water available

Equations for unit operation _i_
- Component k, material balance:

$$\text{Out} - \text{in} = \text{pickup}$$

If ctype is P, then

$$W_i C_{i,k} + \sum_s S_{s,i}(C_{i,k} - \text{cs}_{s,k})$$

$$+ \sum_{j \neq i} X_{j,i}(C_{i,k} - C_{j,k})$$

$$+ \sum_r X_{r,i}(C_{i,k} - C_{r,k}) - L_i C_{i,k} = \text{pickup}_{i,k}$$

- Inlet concentration for component k:

$$\sum_s S_{s,i}\text{cs}_{s,k} + \sum_r X_{r,i}C_{r,k} + \sum_{j \neq i}X_{j,i}C_{j,k} = \left(\sum_s S_{s,i} + W_i + \sum_{j \neq i}X_{j,i} + \sum_r X_{r,i} \right) B_{i,k}$$

If $\text{ctype}_{i,k}$ is P, then

$$B_{i,k} \leq \text{bmax}_{i,k}$$

- Overall water balance:

$$\text{In} \geq \text{out} \qquad (\text{excess goes to treatment})$$

$$\sum_s S_{s,i} + W_i + \sum_{j \neq i}X_{j,i} + \sum_r X_{r,i} \geq \sum_{j \neq i}X_{i,j} + \sum_r X_{i,r} + L_i$$

Equations for regenerator r

- Component k balance: If $\text{rmode}_{r,k}$ is fixed outlet concentration,

$$C_{r,k} = \text{rspec}_{r,k}$$

else if $\text{rmode}_{r,k}$ is fixed percentage removal

$$\left[\frac{1 - \% \text{ removal}_{r,k}}{100} \right] \sum_i C_{i,k} X_{i,r} = \sum_i X_{r,i} C_{r,k}$$

- Overall water balance
  Return to process:

$$\sum_i X_{r,i} = k1_r \sum_i X_{i,r}$$

To treatment:

$$X_{r,T} = k2_r \sum_i X_{i,r}$$

Other constraints

■ Secondary water supply limit:

$$\sum_i S_{s,i} \leq \mathrm{smax}_s$$

Secondary water not used in the process goes to treatment.

Objective functions

■ Minimum fresh water:

$$\sum_i W_i + \sum_r \left[(k1_r + k2_r - 1) \times \sum_i X_{i,r} \right]$$

■ Minimum wastewater:

$$\sum_i W_i + \sum_r \left[(k1_r + k2_r - 1) \times \sum_i X_{i,r} \right] + \sum_s \mathrm{smax}_s - \sum_i L_i$$

References

1. M. E. Goldblatt, K. S. Eble, and J. E. Feathers, "Zero Discharge: What, Why, and How," *Chem. Eng. Prog.*, 89(4): 22–27, April 1993.
2. R. Kelfkens, SASTECH (Pty) Ltd., Secunda, South Africa, personal communication, 1994.
3. A. P. Rossiter, H. D. Spriggs, and H. Klee, Jr., "Apply Process Integration to Waste Minimization," *Chem. Eng. Prog.*, 89(1): 30–36, January 1993.
4. Y. P. Wang and R. Smith, "Wastewater Minimization," *Chem. Eng. Sci.*, 49: 981–1006, 1994.
5. M. Beychok, *Aqueous Wastes from Petroleum and Petrochemical Plants*, Wiley, New York, 1967.
6. R. Smith, University of Manchester Institute of Science and Technology, personal communication, 1994.

18

Ranking Options for Pollution Prevention and Pollution Control Using Graphical Methods*

Alan P. Rossiter, Ph.D., and Jimmy D. Kumana
Linnhoff March, Inc.
Houston, Texas

Michael K. Ozima
Southern California Edison Company
Rosemead, California

Introduction

Previous chapters have discussed how process integration methods can be used to identify good options for reducing emissions in a systematic way. Once the various opportunities have been identified, the next step is to determine the relative costs and benefits of each option. This chapter addresses this issue. It describes a simple numerical and graphical procedure for ranking pollution prevention options. This provides a very clear representation of the relative costs and emission reduction benefits of each option and combination of options.

---

*Adapted with permission from "Rank Pollution Prevention and Control Options," *Chemical Engineering Progress*, February 1994. Copyright 1994, American Institute of Chemical Engineers. All rights reserved.

The methodology can be applied to a wide range of industries, and it is particularly well suited to dealing with options for reducing airborne emissions. It can be used for new designs or retrofits. A company could use it to identify the most cost-effective technical strategy for pollution prevention at its facilities. Businesses could use it as a tool to assist in the commercial trading of emissions offsets (or credits). And regulators could use it to establish a basis for setting realistic *standards of performance,* incorporating both technical feasibility and economics. In this chapter the procedure is illustrated by examples based on NO_x emissions from an oil refinery and volatile organic compound (VOC) emissions in dry cleaning.

Overall Procedure

The steps in the procedure are as follows:

1. List emission sources, emission rates, and applicable pollution prevention and control options

Listing the emission sources and rates establishes a *base case.* Identifying appropriate technologies to reduce the releases typically involves a review of literature sources or use of the approaches described in Part 3 of this book (e.g., transferring emission reduction methods from other applications). Table 18.1 is a list of typical sources of NO_x emissions and technologies for reducing them. These are applicable to a very wide range of process plants.

2. Establish the range of application and the cost relationships for each of the waste minimization or control technologies

Typical values for NO_x control technologies are given in Table 18.2.

Heat integration is an important method for reducing all combustion-related emissions. Pinch analysis (see Part 2 of this book) is the most convenient way to relate heat-transfer area and hence capital cost to process heating requirements, which in turn can be related to combustion-related emission rates (for example, NO_x, SO_x, and particulates). For other control technologies, literature values or engineering databases may be used to determine appropriate relationships.

The relationship between achievable emissions reductions and cost is not always a continuous function. For example, if a certain technology reduces NO_x levels from 100 to 90 ppm at a cost of $100 per ton of NO_x removed, one might assume that the same technique could be used to reduce the level to 85 ppm at an incremental cost of, say, $150 per

TABLE 18.1   NO$_x$ Emission Sources and Control Technologies

a. NO$_x$ Sources	
Source	NO$_x$ rate (kg/MW fired)
Boilers/furnaces	
Fuel gas	0.31
Residual fuel oil	0.67
FCC regenerator/CO boiler	0.46

SOURCE: Ref. 6. These figures include allowances for both *fuel* NO$_x$ and *thermal* NO$_x$. The FCC value includes coke burned in the regenerator and fuel gas fired in the CO boiler, and the literature value has been adjusted to account for this.

b. NO$_x$ Prevention and Control Methods
Conservation measures
Heat integration (HI)
Combustion modifications
Low excess air (LEA) burners
Staged fuel or staged air burners (SB)
Flue gas recirculation (FGR)
Stack gas controls
Selective catalytic reduction (SCR)
Selective noncatalytic reduction (SNCR)

1 kg/MW = 0.646 lb/million Btu.

TABLE 18.2   Costs of NO$_x$ Control Options

Technology	Capital cost ($/MW)	Typical operating cost ($/t; NO$_x$ removed unless otherwise indicated)	NO$_x$ reduction (%)
Heat integration		11,000–25,000 *saving*	10–30
LEA burners	6,550	700–1,700 *saving* (1% fuel saving)	15
Staged burners	6,550	None	55
Flue gas recirculation	12,000	$260/(yr · MW)	75
Selective catalytic reduction	4,440	880	70
Selective noncatalytic reduction	102,000	1200	85

The scope and economics of heat integration are site-specific and are most easily estimated by using pinch analysis.

1 metric ton (t) = 1 Mg = 1.102 U.S. ton.
1 MW = 3.412 million Btu/h.
SOURCE: Refs. 6 and 7.

ton. Such relationships do hold for some technologies, e.g., heat integration. However, most NO_x control technologies are not characterized in this way. Rather, they appear to offer only one level of pollution reduction, with one cost associated with this improvement. Two explanations account for this:

- In some cases (e.g., low-NO_x burners), emission reductions are obtained by a discrete design change. This change has a unique benefit, and no additional benefit can be derived by further expenditure.

- In other cases, a measure of *preoptimization* has been carried out by designers and vendors, and the data quoted simply represent an established economic design point.

In the examples discussed here, the emission reduction technologies are characterized by single operating points, unless clear cost/benefit relationships are readily available.

3. Determine the mutual compatibility of each technology with each of the others

To achieve a given level of emissions, it may be desirable to use two or more emission reduction technologies. However, not all combinations are feasible; and even when a given combination is feasible, the benefits of the technologies may not be directly additive. Four different scenarios are possible:

Mutually incompatible. These are either/or choices. The different types of low-NO_x burners provide good examples of mutual incompatibility—you cannot install more than one type of burner for the same duty.

Directly additive. The effects of some waste minimization or control options are directly additive; i.e., if option A reduces emissions by 100 tons/yr and option B reduces them by 200 tons/yr, the two together reduce emissions by 300 tons/yr. This is generally true only where the options address separate emission sources.

Partially additive. Where two or more emission reduction techniques are applied to the same source, the principle of diminishing returns invariably applies. For example, if burner modifications and stack gas NO_x reduction are both applied to the same furnace, the total decrease in the NO_x rate is less than the sum of the effects obtained with the individual control technologies applied separately. This effect can be approximated as follows:

$$ER = base\left(1 - \frac{x}{100}\right)\left(1 - \frac{y}{100}\right) \qquad (18.1)$$

where ER is the emission rate with two waste minimization or control technologies in place, base is the *uncontrolled* emission rate, and x and y are the percentage reductions of the two separate control techniques when used independently. If three or more technologies are used, Eq. (18.1) is extended by adding additional $(1 - x/100)$-type terms as required.

Mandatory sequence. In a few cases, there is only one possible sequence for installing waste minimization or control techniques. For example, in reducing SO_x emissions in a refinery, a Claus unit can be installed to remove the bulk of the sulfur from refinery fuel gas and hence eliminate most of the SO_x emissions. The Claus unit produces a sulfur-rich *tail gas,* and a tail gas scrubbing unit (e.g., Stretford unit) can be added to remove the sulfur from this stream. However, one would not install a tail gas unit if there were no Claus unit.[1]

4. Calculate the maximum reduction in emissions achievable with each technology and each permissible combination of technologies, and determine the corresponding total cost

This is accomplished by applying the cost and emission relationships established in step 2 and the compatibility rules established in step 3.

Both capital and operating cost contribute to the total cost. To make the comparisons meaningful, it is necessary to convert all costs to a consistent basis. The *total annualized cost* (TAC) is a very convenient basis to use, where

$$TAC = operating\ cost + b \times capital\ cost \qquad (18.2)$$

and b is the *annualization factor.*

5. Establish which technology or combination of technologies provides the least-cost solution for any given reduction in emissions

This can generally be done simply by inspecting the results obtained in step 4. However, when many options have to be evaluated, spreadsheet methods or mathematical programming techniques may be more appropriate.

6. Plot the results (minimum cost against emission rate)

This curve represents the lowest expenditure required to reduce emissions *from the base case* to the specified level, by using the available technologies. Depending on requirements, this plot can take different forms. For example, costs can be in terms of total annualized cost, average control cost, or incremental control cost; and emission rates can be in terms of tons per year, kilograms per second, pounds per barrel, and so on. The most useful representation is probably the minimum average control cost curve. Here the *average control cost* (in dollars per metric ton) is defined as the total annualized cost (in dollars per year) divided by the total reduction in emissions (in metric tons per year); it represents the *average* cost (in dollars per metric ton) for reducing the specified class of emissions from the base case to a given level. The *minimum* average control cost is that obtained when the specified reduction in emission rate is achieved by *using the most cost-effective available technical means*. These costs are plotted versus the emission rate to generate the curves.

Plots of this type provide a very convenient way of representing the cost-effectiveness of pollution reduction technologies. Further, the plots can be used to determine how far emission rates can be reduced subject to specified economic constraints, such as a limit on the cost per ton of pollutant reduction.

An important assumption implicit in the above discussion is that where there are several similar sources of emissions, they are all treated in the same way, i.e., the same technological solutions are applied to all similar emission sources to achieve any required emission rate. If this assumption is relaxed, it introduces greater flexibility and may, in some cases, reduce costs. This is discussed further in the refinery example below.

Application to Refinery NO_x Emissions

Let us consider the ranking of options for reducing NO_x emissions in a typical 100,000 bbl/day (15,900 m³/day) refinery, described in Table 18.3 and Fig. 18.1. An overall (approximate) heat and material balance is developed first, by using in-house data and literature sources.[2,3] The heat balance and the refinery fuel balance establish the base case for combustion-related emissions, which include both *fuel-firing* emissions and *fluid catalytic cracker* (FCC) regeneration emissions. The procedure then continues as follows:

1. List emission sources (with emission rates) and available pollution prevention and control options applicable to each source

The emission sources and some appropriate technologies for address-

TABLE 18.3   Refinery Design Basis

Capacity = 100,000 barrels per stream day (bbl/day) (1 bbl = 0.159 m³)
On-stream factor = 8400 h/yr
Feedstock = blend of 90% Alaskan North Slope and 10% California heavy crudes
Heating requirement
    Total heat absorbed (base case): 287.3 MW (980.4 million Btu/h)
    Furnace and boiler efficiency: 80% (on HHV)
    Fuels:
        Refinery fuel gas (90% of base-case demand)
        Residual fuel oil for remaining duty [0.5% sulfur; $10.24/MWh ($3.00/ million Btu) fired]
Process units
    Crude distillation
    Vacuum distillation
    Naphtha, diesel, and heavy vacuum gas oil hydrotreaters
    Hydrocracker
    Fluid catalytic cracker
    Naphtha reformer
    Alkylation
    Delayed coker

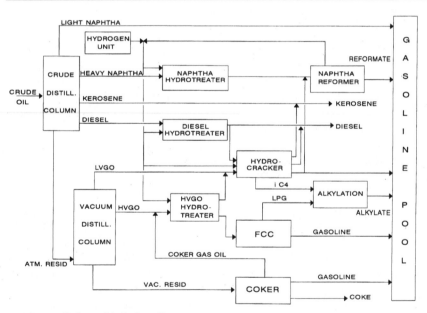

Figure 18.1   Refinery block flow diagram.

ing refinery NO_x emissions are listed in Table 18.1. Base-case emission
rates from the boilers and furnaces in the refinery amount to approximately 1148 U.S. tons per year (tons/yr) [1041 metric tons per year
(t/yr)], and from the FCC regenerator/CO boiler system, emissions are
938 tons/yr (851 t/yr), giving a combined total of 2086 tons/yr (1892
t/yr). All boilers and furnaces are considered equivalent in this analysis (but see the discussion under "Optimum Control Strategies").

2. Establish the range of application and the cost relationships for each of the waste minimization or control technologies

The attainable emission reductions and applicable cost data for NO_x control options are given in Table 18.2. In the case of the heat integration option, a sitewide pinch analysis was carried out for the refinery, to establish the impact of heat integration on heat demand and to determine the capital versus energy cost tradeoff for the site. This was based on the data in Table 18.3, together with a capital cost for heat-transfer area of $100 per square foot ($1076 per square meter). The analysis showed that fuel savings through improved heat integration offered very favorable economics for all practical levels of heat integration. The maximum practical scope for heat integration is limited by the need to burn all the fuel gas produced at the site.

3. Determine the mutual compatibility of each of the technologies with each of the others

The options for reducing NO_x emissions fall into three basic categories: conservation measures (e.g., reducing energy needs in order to reduce fuel firing), combustion modifications (e.g., using different burner types), and stack gas controls (i.e., end-of-pipe elimination of NO_x).

The options within each category are generally mutually incompatible. For each boiler and furnace, it is therefore possible to use one combustion modification technology in combination with one stack gas control technology. The benefits would be partially additive, as previously discussed. [Options combining two end-of-pipe options, such as *selective catalytic reduction* (SCR) and *selective noncatalytic reduction* (SNCR), are not considered here, although they are possible under certain circumstances.[4]]

If the total firing requirement in boilers and furnaces can be reduced by heat integration alongside burner modifications and/or stack gas controls, this reduces NO_x emissions on a *partially additive* basis. However, the maximum practical scope for heat integration is limited by the need to burn all the fuel gas produced by the refinery (see Table 18.3). If the process energy requirement falls below the heat available from refinery fuel gas (90 percent of the base-case fuel requirement), the fuel gas will become surplus. Unless there were some way to use this surplus gas beneficially (e.g., power generation or export to a neighboring site, with sale of NO_x offset credits), it would have to be flared, and any further heat integration would not lead to a reduction in total fuel firing on the refinery. This is the situation at this refinery, and it sets a practical limit on the scope for useful heat integration.

For the FCC regenerator, burner modifications and heat integration are not feasible options. Only the stack gas control options (SCR and SNCR) are viable, and there are no feasible combined options. Any reductions in NO_x emissions in the FCC/CO area would therefore be directly additive with improvements elsewhere.

4. Calculate maximum emission reductions and corresponding total costs

The total reduction in NO_x emission rate and the corresponding TAC is calculated *for every feasible option and combination of options.* Although this sounds daunting, in practice, the total number of feasible combinations is small—11 for the boilers and furnaces and two for the FCC/CO boiler area—and this exercise is readily handled with a spreadsheet.

Table 18.2 lists the maximum reduction in NO_x emissions for each of the individual technologies, together with data on the associated capital and operating costs. The TAC is then calculated from Eq. (18.2). In this example, a value of 0.27 is used for the annualization factor b. This is equivalent to requiring a 15 percent rate of return with a 5-year plant life.

Heat integration can achieve a 10 percent reduction in fuel requirements. The fuel eliminated—*residual fuel oil* (RFO)—is a relatively high NO_x fuel, and the actual percentage reduction in NO_x production from boilers and furnaces is roughly 19 percent (relative to the base case). There is no impact on coke burning in the FCC regenerator.

The performance for options that comprise two or more compatible technical solutions in the boilers and furnaces is assessed by applying the partially additive formula, Eq. (18.1). For example, if the maximum heat integration is carried out and staged low-NO_x burners are provided for all boilers and furnaces, the overall NO_x emission rate from boilers and furnaces will be

$$ER = 1148 \left(1 - \frac{19}{100}\right)\left(1 - \frac{55}{100}\right) = 418.4 \text{ tons/yr (379.6 t/yr)}$$

The cost for this combined option is the sum of the TAC for maximum heat integration (− $2.496 million per year, a negative number, reflecting a very favorable return on investment) and the TAC for staged burners ($0.572 million per year). The overall TAC is therefore − $1.924 million per year.

The NO_x elimination in the FCC/CO boiler area is essentially a different problem, separate from NO_x in the boilers and furnaces. The two stack gas controls—SCR and SNCR—are the only two (mutually exclusive) options considered. Reductions in emissions from the FCC/CO boiler area and those obtained in the boilers and furnaces are additive when we calculate overall reductions in refinery emissions.

5. Establish least-cost solutions for any given reduction in emissions

By using the results from step 4, it is a simple matter to tabulate the TAC and overall refinery NO_x emission rate for all feasible options and combinations of options (remembering that the FCC/CO boiler emissions are directly additive with the boiler and furnace emissions). It is then possible, by inspection, to identify and tabulate the least-cost option (or combination) for reducing overall emissions to (or below) any given level.

6. Plot the results (minimum cost versus emission rate)

The table of least-cost options generated in step 5 can be plotted directly as a minimum TAC versus emission rate (Fig. 18.2). Dividing the TAC values by the reduction in NO_x emissions (compared to the base case) gives the minimum average control cost plot (Fig. 18.3). An *incremental control cost plot* (Fig. 18.4) can be developed by dividing the difference in TAC between any two consecutive data points in Fig. 18.2 by the corresponding reduction in NO_x emissions. This plot is useful in highlighting the economic implications of individual increments of pollution reduction.

Optimum control strategies

The base-case refinery has NO_x emissions of 2086 tons/yr (1892 t/yr, or 0.12 lb/bbl). The first 11 percent of these emissions can be eliminated

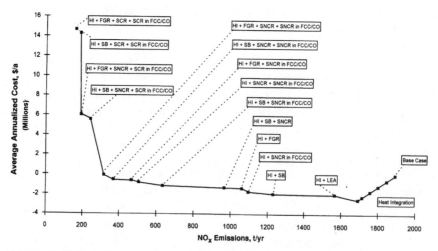

Figure 18.2   Refinery NO_x emissions: total annualized costs.

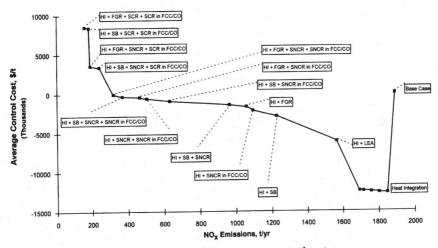

Figure 18.3 Refinery NO_x emissions: minimum average control costs.

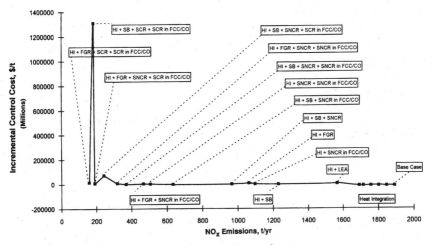

Figure 18.4 Refinery NO_x emissions: incremental control costs.

at a profit by heat integration, and the minimum average control cost is therefore negative (see Fig. 18.3). After heat integration has been maximized, it is necessary to install additional equipment to obtain further emission reductions. Each equipment addition increases the minimum average control cost—i.e., reduces the profit—and ultimately turns it into a loss.

Low excess air (LEA) burners, applied to boilers and furnaces, can achieve a reduction of 15 percent in NO_x emissions from these sources, and they offer the next increment of minimum-cost improvement after heat integration. However, if larger NO_x reductions are needed, LEA burners *should not* be installed. It becomes more cost-effective to install staged burners (a competing technology) in the boilers and furnaces. Together with the heat integration, this reduces the emissions to 1355 tons/yr (1229 t/yr).

If the required NO_x rate is between 1355 and 1208 tons/yr (1229 and 1096 t/yr), the minimum-cost solution involves heat integration plus *SNCR in the FCC/CO boiler area.* In this range of emissions, staged burners *are not* a part of the optimum solution. This highlights an important fact about the control cost curves: Each data point represents a unique set of technical options that define the minimum-cost means of achieving the specified emission rate. These options may or may not include elements present in the adjacent data points. The plots therefore provide guidance in developing minimum-cost strategies to reduce emissions *from the base case* down to any selected value. However, if some of the control strategies are *already* implemented, a new base case is created, and the same strategies do not necessarily apply.

Further reductions in NO_x emissions require different—and increasingly expensive—combinations of control technologies. The maximum reduction in NO_x emissions using the available technologies is 91.6 percent, yielding an emission rate of 175.5 tons/yr (159.2 t/yr). However, it is significant that the average control cost is negative for reductions of up to 80.5 percent [406.8 tons/yr (369.0 t/yr) released], due to the savings achieved by heat integration. Beyond this point it becomes necessary to incorporate SCR into the control strategy for the FCC/CO boiler, and this causes a very steep rise in costs. This is very clear from Fig. 18.4 (the incremental control cost plot), which is totally dominated by the two "spikes" at the low-emission-rate end. These correspond to the addition of SCR in the FCC/CO boiler area and in the furnaces and boilers. In effect, the final 231.3 tons/yr (209.8 t/yr) of NO_x elimination costs nearly $15 million per year, or $65,000 per ton ($72,000 per metric ton). Addition of SCR to the boilers and furnaces actually costs nearly $1.9 million per ton ($2.1 million per metric ton) of NO_x removed. This compares with the market rate for NO_x offsets, which is in the range of $27,000 to $33,000 per metric ton. However, some current legislation *mandates* the use of SCR in certain applications.

Treatment of individual emission sources

The form of the emission control cost–emission rate curves depends to some extent on how many emission sources are being considered and whether it is permissible to implement different control technologies at similar NO_x sources. The cost/benefit curve for a single emission source has a strictly limited number of possible points, dictated by the number of available waste minimization and control technologies and feasible combinations. However, as the number of separate sources increases, the number of points also increases; an infinite number of sources would yield a continuum.

All the plots considered thus far have assumed that the same control technology(ies) is (are) applied at all similar emission sources. This results in a plot very similar to what would be obtained for a single emission source. However, if the use of different options at each source is allowed, the result is significantly different. This is illustrated in Fig. 18.5. (Note that the base case has been modified to include the maximum practical heat integration. The remaining control costs are all positive, which greatly simplifies the graph.)

In Fig. 18.5, it is assumed that the controls can be added to *each* of the furnaces or boilers *one at a time*. It is further assumed that there is an arbitrarily large number of these items. The result is a "smoothed" curve, showing increasing average control cost as the emission rate falls. Furthermore, some options that previously represented minimum-cost solutions for certain emission rates are no longer viable, for example, LEA burners.

All the results in this example are based on the published cost data in Table 18.2, and it is assumed that the pollution controls are added as retrofits to an existing refinery. However, similar relationships can be developed for the incremental costs of incorporating the same technologies in new plant designs. For any practical study, current tech-

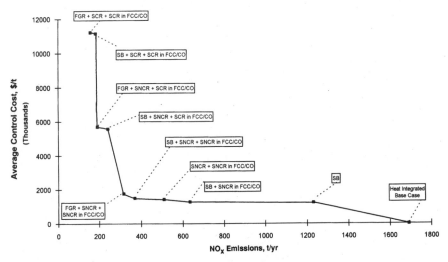

Figure 18.5 Refinery NO$_x$ emissions: smoothed control costs.

nical and cost data should be used to develop appropriate comparisons. Technical developments, e.g., improvements in catalytic reduction, could significantly change the economics of the various control options.

Application to Dry Cleaning Emissions

The ranking procedure can be applied in many different situations. In addition to addressing NO_x emissions as described in the example above, it has been applied to SO_x and VOC emissions and with some modifications also to aqueous emissions.[1] The technique has been used for large-scale processes (the refinery example given above), and it has also been applied to small-scale operations such as dry cleaning.[5]

The dry-cleaning process is illustrated schematically in Fig. 18.6. The major steps in the process are analogous to those of conventional laundering, except that nonaqueous solvents are used. Losses of this solvent give rise to VOC emissions. The major sources of solvent losses are

- Solvent retained in filter elements and muck
- Distillation bottoms

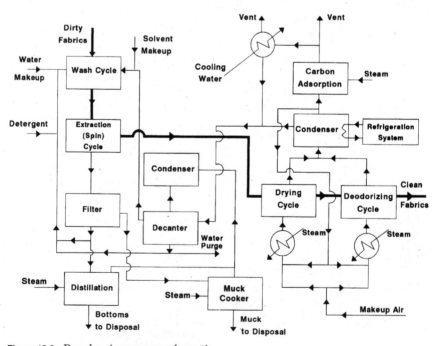

Figure 18.6 Dry-cleaning process schematic.

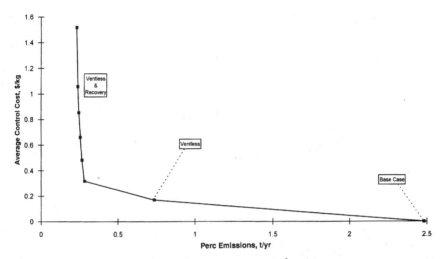

Figure 18.7 Dry-cleaning study: minimum average control costs.

- Dryer exhaust
- Fugitive emissions

The results of the dry-cleaning study are illustrated in Fig. 18.7. Here the minimum average control cost is plotted as a function of losses of solvent (perchloroethylene, commonly termed *perc*) from a dry-to-dry dry-cleaning machine. Three add-on options for reducing emissions are considered:

1. Refrigerated condensing—vented
2. Refrigerated condensing—ventless
3. Filter cartridge stripping (recovery)

Figure 18.7 shows the minimum net cost of environmental control for any given level of perchloroethylene emissions. It relates the *control cost* (in dollars per kilogram of reduction of perchloroethylene emissions below base-case levels; 1 kg = 2.2046 lb) to the emission rate (in metric tons per year; 1 t/yr = 2204.6 lb/yr) and identifies which technology or combination of technologies gives the minimum-cost solution.

From Fig. 18.7 it is apparent that the most cost-effective first step in emission control is the installation of a ventless condenser. This gives a control cost of $0.170 per kilogram and offers a lower emission rate than the addition of a recovery unit. Vented condensing is never competitive with these options: Its total annualized cost (including the cost of the perchloroethylene lost in the vented stream) is higher than that of the ventless system. The optimum locus therefore goes from the base case directly to ventless condensing, and then to ventless condensing plus filter recovery. The "tail" on the left-hand side of Fig. 18.7 represents the effect of lowering the temperature of the ventless condenser. This reduces emissions slightly, but increases costs significantly.

Implications of Case Study Results

The case studies illustrate a number of points, many of which are widely applicable. These are the three most important:

- Heat integration offers the most cost-effective means of reducing NO_x emissions. It also reduces other greenhouse gases and particulates, and it can significantly reduce wastewater discharge rates (see Part 2 of this book).

- Other conservation-based techniques, such as ventless dry cleaning, also offer attractive economics. Published results using this technique[1] have also demonstrated the favorable economics of several other conservation-based measures. These include water recycle to reduce wastewater discharges, Claus sulfur recovery for SO_x emissions, and improved tank seals and leak detection and repair programs to minimize certain types of VOC emissions.

- Many end-of-pipe control methods are disproportionately expensive.

These observations are very similar to those obtained by taking other systematic approaches to pollution prevention discussed elsewhere in this book (see, for example, Chap. 21).

Benefits of the Numerical/Graphical Approach

The major benefits of this approach are as follows:

- It provides a generic methodology for evaluating and ranking available industrial waste minimization and pollution control technologies.
- Plots of the results obtained by using the procedure give a very clear, concise, and easily understood representation of the economic implications of environmental compliance options.
- It introduces a rational basis for establishing *standards of performance* for each industry or each manufacturing site.
- It is consistent with the concept of *market-based* environmental legislation.

References

1. A. P. Rossiter, H. Klee, Jr., and R. D. Sung, "Process Integration Techniques for Pollution Prevention," American Chemical Society I&EC Special Symposium on Emerging Technologies for Hazardous Waste Management, Atlanta, September 1992.
2. J. H. Gary and G. E. Handwerk, *Petroleum Refining—Technology and Economics*, 2d ed., Marcel Dekker, New York, 1984.
3. M. Beychok, *Aqueous Wastes from Petroleum and Petrochemical Plants*, Wiley, New York, 1967.
4. D. A. Rundstrom and J. L. Reese, "Catalyst Air Heater Retrofit Reduces NO_x Emissions," *Power Eng.* 96(8): 38–40, August 1992.
5. R. D. Sung, A. P. Rossiter, and J. D. Kumana, "Process Integration and the Environment," International Energy Agency Workshop on Process Integration, Gothenberg, Sweden, January 28–29, 1992.
6. J. E. Mincy, "Controlling NO_x to Obtain Offsets or Meet Compliance," 14th Annual Industrial Energy Technology Conference, Houston, April 1992.
7. R. McInnes and M. van Wormer, "Cleaning Up NO_x Emissions," *Chem. Eng.*, 97(9):130–135, September 1990.

Putting It
into Action

19

Managing the Design Process

Darryl W. Hertz
Manager, Pollution Prevention Program
The M. W. Kellogg Company
Houston, Texas

Introduction

Part 1 of this book looked at a number of the issues—economic, practical, legal, and technical—that underlie current environmental trends, and Parts 2 through 4 described the major systematic technical approaches currently available for pollution prevention. The types of information provided in these early sections are necessary for successful implementation of waste reduction programs—but they are not sufficient. To "make things happen" by developing and implementing good "clean" process design improvements, it is also necessary to provide the right organization and work process environment. This is the theme of this final part of the book. This chapter specifically describes systematic methods for addressing organizational and technical problems encountered while implementing sustainable pollution prevention process improvements. Chapter 20 introduces a standardized accounting system that can be used as an environmental management tool. The remaining chapters (21 through 23) describe a variety of pollution prevention programs that have been successfully implemented by industry and by government agencies.

Throughout this chapter, discussions of key program concepts and issues are presented that should be considered by managers of pollution prevention programs in all types of organizations—producers of chemicals and other products, engineering and construction companies, electric utilities, engineering consultants, government agencies, and others. We start by considering the differences between pollution prevention engineering and traditional engineering, and how pollution prevention affects the work process. Next we review the need for new design tools. Then we look at the various components needed in a corporate pollution prevention program and how pollution prevention options are initiated, developed, and implemented. In practice, there are a number of barriers to effective implementation, and these are also discussed. Finally, verification of pollution prevention results and continuous process improvement are addressed.

Comparison of Pollution Prevention Engineering with Traditional Process Engineering

Pollution prevention engineering, or the systematic incorporation of environmental considerations into the production process, is very different from traditional process design efforts. New perspectives are demanded by ever-increasing environmental constraints placed on industry. To meet this demand successfully, more people with a wider range of expertise are needed. Also, they are needed throughout the process design rather than at one or two stages where they traditionally hand off the design to others in other groups. The resulting designs must not merely conform to environmental regulations today, but must anticipate environmental restrictions in the reasonable future. These restrictions typically do not improve process profitability and may be considered parasitic on what would otherwise be a more competitive process.

Developing and implementing pollution prevention process improvements are far more demanding than traditional process design efforts. They require significant knowledge of the entire plant design, plant operations, maintenance procedures, and applicable and potentially applicable environmental regulations. This means they demand far more than energy optimization process improvements such as pinch analyses. Pollution prevention is *not* the same old process optimization that engineers have always been doing. The environmental restrictions relevant today push the limits of existing design tools and technologies. The restrictions often demand totally new approaches to attain the needed goals.

Pollution Prevention and the Design Work Process

The design work process is the same for any production or manufacturing facility. Each stage (see Fig. 19.1) of process design increases design confidence, accuracy, and relevant systems information from the initial research stage to the construction and start-up stage. The highest potential for pollution prevention exists in the first stage. However, this activity usually occurs in separate long-term programs focused on development of fundamentally new production or manufacturing processes, and a significant lead time is required to bring pollution prevention concepts from this stage to full implementation. For revamp work, which is by far the most common process design activity in the United States today, basic research is rarely involved, unless a new component of the production technology is the intent of the revamp. Most process design activity in the United States today begins with the process synthesis stage of design. This sets the scene for the conceptual design effort, which is often performed by separate engineering and construction companies or similar groups within large producing or manufacturing companies.

It is helpful to distinguish between long-term and short-term pollution prevention options. We define *long-term options* as those process changes that either satisfy environmental requirements expected in the next 5 to 10 years or will take at least 5 years to implement. *Short-term options* are those that can take effect within 5 years. In general, the highest *long-term* pollution prevention potential exists at the earliest stages of process design, and the highest *short-term* potential for

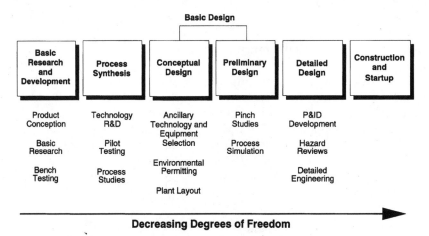

Figure 19.1   Traditional process design work flow.

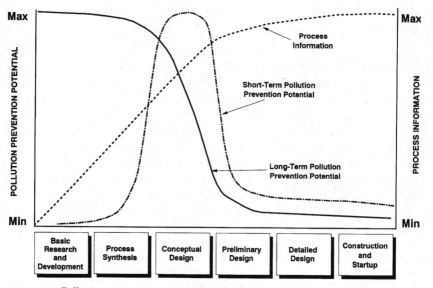

Figure 19.2   Pollution prevention potential in the design process.

implementation of effective pollution prevention exists in the conceptual design stage. This premise is depicted in Fig. 19.2.

In practice, there are usually different groups of designers performing the different stages of process design. It is also highly likely that different people performed basic research, laboratory testing, and pilot testing before initiating the implementation phase of the effort at the conceptual design stage. These groups may be separated geographically and ideologically. This means that different people and groups of people have the ability to do pollution prevention at different stages in the process design work flow, but with different motivations, experiences, goals, budgets, and time frame for completion of the work (see Fig. 19.3).

These differences make it difficult to maintain a consistent philosophy for pollution prevention throughout the design process. To overcome this, the best place to do pollution prevention is in an independent technology development or improvement program or project.[1] Here, a neutral location can be selected where the best chances exist for developing new ideas and that includes experts from a variety of groups such as plant operations, maintenance, and plant engineering, together with those whose primary responsibility is design.

Basic research and development

The first stage of process design is performed as part of a long-term program to identify significantly advantageous new processes for an

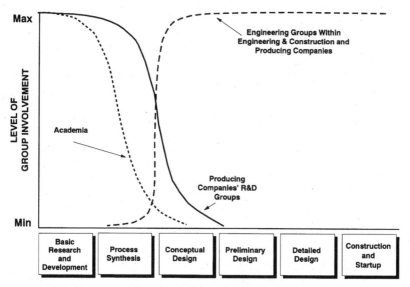

Figure 19.3   Different groups involved in pollution prevention.

existing mature market or to identify new processes for new products for an anticipated new market. This work can be done privately or in some collaboration with universities to leverage limited financial resources or to make use of unique research capabilities. Today, much of this activity involves identifying new production process pathways so that products can be made economically with a much smaller "environmental footprint." This offers the greatest potential for pollution prevention.

Process synthesis

During the process synthesis stage of design, researchers and engineers are generally given few constraints and a significant amount of time to accomplish the best overall process design to produce the needed product for the intended market. Here, exact product specifications are chosen, and the production technology is selected. This often requires specific process studies to be performed before crucial decisions can be made. If the process is a new one where significant questions remain, pilot testing will be performed to define the scale-up parameters.

In the past, the issue of waste generation and management had little impact on the technology selected. Today, however, that is no longer the case. Incorporating pollution prevention at the process synthesis stage offers one of the highest potentials for practical reduction of waste generation.

Conceptual design

In the conceptual design stage, the designer is asked to come up with a process that is viable both technically and economically; but this must be done in the time set forth by strategic planners in the company or in a client's company. These time constraints often reduce the possibility of reviewing ancillary technology and equipment selections. This stage of process design locks in about 90 percent of the design decisions made in the entire design work process and profoundly affects the engineering effort in later design stages. If pollution prevention has not been done in a previous program with fewer time constraints, it will receive little consideration here. This is why a separate program is so important.

This stage is often where grass-roots or revamp projects are kicked off, and significant installation funds are on the table. Contracts are let to the successful bidders, and engineering and construction work begins in earnest. Here, the environmental permitting effort begins for obtaining complex permits prior to construction and start-up. It is done very often with little process information and relies on estimation methods put forth by federal and state environmental regulations. This is also where a previously completed pollution prevention program effort can be of great value because of the information already gathered. The best way to estimate what the process effluents will be and their most likely chemical constituents is to use the combined expertise required for a pollution prevention effort. It is much better to have the environmental engineering designers prepare permit applications based on the best guesses of a wide group of process experts than based on federal and state calculation estimates. Also, the evaluation record can be of great value in explaining why certain process technologies were selected.

Preliminary design

Some combine this stage with the conceptual design stage and call it *basic design*. However, it should be considered a separate stage of design because it is here that the process optimization studies are performed, after most of the process has been defined in the conceptual stage. Decisions have already been made concerning the type of technology for separation, purification, utilities, heat exchange, and end-of-pipe treatment. Here process simulators can provide greater accuracy about what can be expected regarding the production process itself. However, they are still very deficient in predicting chemical species in dilute process effluent or waste streams. Also they are deficient in simulating newly commercialized separation technologies and any new technology for that matter. They do their best work with established and understood technologies critical to production portions of the plant.

Detailed or equipment design

This stage of design is very well understood and involves the greatest overall engineering effort by far. Here, process piping and instrumentation diagrams (P&IDs) are developed, and detailed equipment designs are completed. This stage includes *design hazard reviews* in which the safety and operational aspects of the design are considered and approved. Aspects of maintainability, operability, constructability, and reliability are confirmed here. For pollution prevention, little is possible in this stage of design. Having said that, we stress that waste reduction must be understood as important, or else waste generation issues and environmental release issues will be not be balanced with the safety, maintainability, and operability issues considered. Consultation with the people who carried out the pollution prevention program is important at this stage, to ensure that the correct balance is maintained.

Construction and start-up

Once construction begins, a new group with different priorities takes over. They are involved with design engineering only as necessary to ensure that a safe and correct plant or process is built. Pollution prevention considerations here are distinctly similar to those in an operating plant. They are also similar to those required in a hazardous waste site remediation operation. Here, waste management and spill prevention are very important. Techniques used emphasize the health and safety of the construction staff, as they should. Here, operability and maintainability issues give way to constructability issues. The new group assumes that the design engineers have already thought operability and maintainability issues through. This stage of design has the lowest potential for pollution prevention in terms of the operational aspects of the process.

In the start-up phase, the first glimpses of operational and environmental issues become apparent—often quite dramatically. Here, the reality of the process design work becomes evident, and the results of the pilot testing are confirmed. Also, the environmental releases and effluents are confirmed. The consequences of exceeding the permitted releases to the environment are severe. They usually result in significant delays in "coming on-line" and may involve last-minute installation of expensive end-of-pipe treatment systems not included in the early design stages. This is when the value of the pollution prevention programs in predicting such effluents and releases becomes most evident.

The various technical approaches described in the earlier parts of this book can be correlated with the different stages in the design process. The knowledge-based approaches presented in Part 3 are most directly applicable to the process synthesis stage, where basic technology selection decisions are made. Pinch analysis (Part 2) and numerical optimization (Part 4) relate most closely to the conceptual design stage, where technology selections are finalized, and the preliminary design stage, where designs are optimized.

The Need for New Pollution Prevention Design Tools

If environmental considerations are to become an integral part of every process design work flow, new design tools and information systems are needed to make this happen. The core of this book (Parts 2, 3, and 4) describes a number of tools that have been developed in recent years. New information tools and design tools are needed to allow design teams to understand the environmental implications of their designs as they are creating them, providing "expert advice" as and when needed, and increasing the engineers' confidence in the final design. If new and quickly obtainable process technology information is not available, engineers will simply use what has worked well in the past for the same application.

One important key to accomplishing waste reduction is the incorporation of relevant information into conceptual design decisions through the development of a framework that allows the designer to examine design alternatives rapidly. There is a huge and rapidly growing body of knowledge to consider. Many current research projects are identifying individual causes of pollution and viable pollution prevention solutions in process facilities. This work requires a broad range of information including an understanding of the interactions of pollutants and contaminants with product streams, catalysts, and equipment, as well as their fate and transport if discharged. Design data for the hundreds of already demonstrated or proposed pollution prevention technologies and methodologies are also required. Unfortunately, systems for obtaining much of this information either are not widely accessible or do not exist.[2]

Chapter 11 describes one approach to this problem—a unit operations–based pollution prevention solution database. Another, more comprehensive system under development through a joint industry/university/government program is called the Clean Process Advisory System (CPAS).[3] This system includes design tools containing new technology performance data; experience-based information or expert guidance; usable health, environmental, safety, and risk data; and federal, state, and local regulatory information.

Components of a Corporate Pollution Prevention Program

This section discusses the various components or programs in almost all corporate pollution prevention programs. Each program type addresses several stages of the process design work flow discussed previously (see Fig. 19.4). The resources available, time, budget, personnel, and focus vary considerably. A structured program is essential to maximize program benefits, and it must address the real driving forces for the company performing the pollution prevention program. Some address short-term issues, and others deal with issues potentially valid in 5 to 10 years. All programs must look beyond current environmental compliance needs. Compliance is the bare minimum.

To overcome limitations such as budgets, available facilities, and staff experience, it has become very popular to pursue programs collaboratively to leverage resources and to strive for a "win-win" situation between companies. Today, through industrial consortia such as the Center for Waste Reduction Technologies (CWRT) of the American Institute of Chemical Engineers (AIChE) and the Petroleum Environmental Research Forum (PERF), even competing companies can jointly achieve leveraged programs for waste reduction.

Basic research and development program

This type of program can be considered the same type of effort as the first stage of process design and has the highest potential for pollution prevention to be implemented. Very long-term considerations are pos-

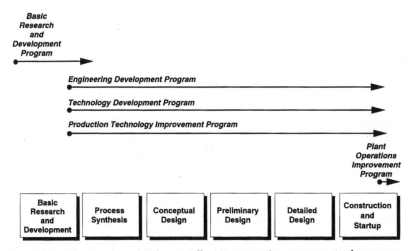

Figure 19.4 Relationship of different pollution prevention programs to the process design work flow.

sible with sizable budgets and staff necessary. Academic involvement is common, to make use of unique expertise and facilities or to leverage corporate capabilities and financial resources. This program makes use of far more expertise than the traditional research and development effort. Additional expertise includes experience from plant operations, conceptual design engineering, detailed design engineering, environmental engineering, and maintenance.

Technology development program

These programs often deal with production technology development or that technology that the company considers its core technology or technological area. This type of program has a very high potential for pollution prevention implementation if the R&D, plant operations, maintenance, engineering, and environmental groups are involved. These programs often have a long-term focus with moderately large staff availability and budget needed. Because these technologies most likely answer product market pressures, they require the greatest need for understanding of the market for the production technology and the product produced from it.

While this type of program has a very high potential for reducing waste generation within the production process, there is often tremendous pressure to spend very scarce R&D money to obtain the maximum short-term return on the investment. This means that the greatest emphasis will be placed on completing a reliable and differentiated process that will make it to commercial production the fastest. Here, pollution prevention will be implemented only if it is an integral part of the development process. It cannot be an add-on consideration. If it is, it will often suffer from not presenting the adequate incremental economics to survive the evaluation and optimization process that the production technology will inevitably go through.

Production technology improvement program

This program type is different from the previously discussed programs in that it focuses primarily on ancillary systems improvement, i.e., processes and systems not directly involved in producing the product for which the overall process was installed. These programs often have short to medium time frames with a relatively small staff and budget available. The program works with established and usually operating processes and encompasses all process design stages beyond basic research and development (see Fig. 19.4).

Following preliminary design, this program is often melded with the installation design effort, and the two become indistinguishable. However, the work done to this point has drawn on a much larger group of experts than a typical process design effort might have. Also, the experts have been drawn from and separated from their normal daily duties to participate in the pollution prevention effort, free from distraction. This makes the overt support of upper management crucial. If pollution prevention is not set out as a corporate priority, identified goals will be unattainable because resources and support will be diverted to other, more traditional programs.

These pollution prevention programs should be free to include all aspects of waste reduction (i.e., source elimination through end-of-pipe treatment), to select the most promising for both the short and long term. In a properly planned program, most groups within the company should be represented. This results in a wider acceptance of proposed process enhancements, operations changes, and capital investments. This program can be the best forum for including current and planned process changes whereby pollution prevention options can be intertwined to produce the best possible economic scenarios.

Engineering development program

There is a large *information gap* between industries, between companies within the same industry, within companies spanning many industries, within large corporations, and even in small companies. This dearth of pertinent information and of the guidance techniques to use it becomes quite evident during pollution prevention programs. For example, questions are often asked that challenge existing paradigms and seek to unearth the true reasons, or lack thereof, for handling process effluent streams as has been done in the past. Answers are not always readily forthcoming.

This creates a tremendous technology information transfer issue. Engineering development programs address this problem. They seek to fill the significant information needs of process designers in the early stages of the process design work flow with tools that make process design more accurate and the work process more efficient. These new tools were discussed earlier in this chapter. I believe they are the only way that pollution prevention issues and challenges can be successfully incorporated into the traditional process design work flow.

With these new tools, new technologies can be introduced more rapidly, because there will be information about others who have already done it. New information tools will reduce perceived risk about introducing new processes or process modifications by making use of previous design information. If this is done correctly, future process design efforts will be improved and/or shortened. The goal is to find the "nugget" of information required for the work at hand. This nugget of hard-to-find information has been nearly impossible to locate with existing engineering design tools.

Operating plant or process improvement program

These programs have been responsible for the highest return on investment for pollution prevention efforts. Often, little or no investment has been needed to implement recommendations of these programs. Their results usually have very good impacts on the operating plant. This is mostly due to the application of the plant or process operators' knowledge of cause and effect. Experience has proved this fact many times over (e.g., see Chap. 22). In such programs, simple process and operating changes that can be implemented quickly are exploited for the improvement of the overall process.

This vast, available repository of process operations information should not be restricted to these programs only. Knowledge of the *true* process is usually the key to successful process modifications for all the pollution prevention programs discussed above. Information exchange (technology transfer) about actual plant experience and system behavior can greatly enhance the effectiveness of design engineers. This information exchange is also important in the engineering development efforts mentioned above. The engineering tools that are developed in these programs to provide information and guidance in pollution prevention process design must include the knowledge base of process operators as well as that of the engineers and researchers.

Initiation and Development of Pollution Prevention Options

For any successful pollution prevention program, a defined program structure must be developed. Further, it must be allowed to evolve over time to adapt to changing corporate pressures, strategic plans, changing markets, and resources available to the program effort. Below are some important considerations for such programs. They may not all apply to every program, but many aspects of those presented will.

Selecting the best team

To get the most out of a creative pollution prevention program, evaluation and development of waste reduction process improvements must be done away from the day-to-day pressures and duties of the participants. This means a separate budget and separate staff are necessary.[1] The best situation occurs when operations, maintenance, process engineering, and environmental engineering are gathered for pollution prevention work. Significant experience must be gathered in one location, often in the face of the demands of regular day-to-day duties. This is where the support of upper management comes into play.

Don't try to generate waste reduction process improvements in just a few meetings. First, you will wear down the participants. Second, they will not have time available to attend the long meetings. Third, they will not have time to shift from their daily pressures to the more "blue sky" environment needed for pollution prevention design generation. Fourth, they will not have time to get to know each other. If they believe that any idea will fly in this arena, then good things will happen—they will begin to share their best ideas that no one has been willing to listen to in the past.

Another important issue should be considered. The group gathered for this purpose should not be accompanied by their supervisors.[4] If they are, they will not be open with their "pet" waste reduction ideas.

Identify program limitations

Within every company there are factors that limit pollution prevention efforts. Many of these limitations can be overcome if the manager of the program can first understand their existence, why they exist, who is responsible for their existence, and, most importantly, who and what resources will be needed to overcome them.

The manager of the effort must find out who is needed to develop and evaluate the designs and if and when that person will be available. It is important that the program sessions be led by someone who is a strong personality with a clear idea of where everyone is going. The available budget to achieve what will be recommended should be explained to all. This will allow realistic goals and deliverables to be formed. The team should understand the time frame for implementing pollution prevention options, as this can significantly influence the desired deliverables.

All those involved should have a valid reason for being involved. Part of their involvement must include some kind of benefit from environmental process improvements. If they have none, they often become impediments to the outcome of the program. If there is a true win-win situation for participants, all involved will strive for the most attractive overall result and thus improve the likelihood of management approval.

The option development methodology

Initiation and development of pollution prevention process improvements or options must be systematic, as discussed in the previous subsection and the next. The methodology must lead to recognizable process modifications in an appropriate level of detail. It must be adaptable to those people available to do the work. Not all the best people can be made available to do this work at all times. There should be significant time between initial idea generation and completion of deliverables. Remember, the best designs are usually a collection of evaluations and improvements on the original idea.

Prioritization of options

The primary purpose of rigorous prioritization of the waste reduction options is to identify the best and most effective designs for preferred implementation. An important secondary purpose is the justification of the selection process to those not involved with the development of the design improvements. This often includes those in management, third parties wanting to collaborate on the selected best process improvements, and those in other parts of the corporation who may want to consider implementation of the same process improvement for the same reasons you are recommending one improvement over the others.

Value of a structured protocol. A structured protocol is necessary for presenting process changes to a very skeptical or risk-averse management with significant short-term pressures and very limited budgets.[5] Managers often have little time to dwell on long-term and/or "soft" considerations not based directly on measurable financial criteria. The bottom line is almost always a short-term bottom line. To put it another way, the real world always intrudes and demands that the best financial decision be made after all aspects of a design change have been weighed. Here is where a rigorous ranking protocol pays off. Evaluation of pollution prevention design improvements must be as rigorous as that of traditional design improvements.

Benefits of a rigorous prioritization. Iron-clad evaluation criteria are needed to rank waste reduction options. Accuracy is not the only benefit. Rigorous analysis also pays dividends when the time comes to explain the thought process of the team in prioritizing pollution prevention options. This exercise also serves another very important purpose—it further scrutinizes the design options from several new angles not previously considered by the design team. The result is a more accurate picture of the process change and the potential issues that will arise when time comes for installation, start-up, and maintenance. The result is a better design with fewer unexpected flaws. It also has a most unexpected and valuable function: It helps the team experts become better experts. Here, they discuss, often heatedly, the relative attributes of each change from many points of expertise not heard before. They learn from other experts in a way not easily done before, as there are generally few opportunities to discuss such material on day-to-day jobs.

Presentation of options to management

Clear and effective verbal and written presentation of new ideas to management is crucial. Format and level of detail are important to the success of the presentation and thus ultimately to the acceptance of pollution prevention recommendations by management. Standards for these aspects of the presentation are suggested below.

Report format. Pollution prevention process improvements should be presented in the correct format, by the correct people, with correct detail, and with the best possible correlation to other ongoing or planned engineering activities. Other improvements in the process that could further improve the overall design should be mentioned. Also, those process changes that could make the design improvement either moot or uneconomical should be discussed. To that end, in the brief descriptions of waste reduction options, other related options and those that could be combined into larger and more integrated process improvements need to be considered. It is not uncommon for design changes of greater significance to lead to improved incremental economics merely due to the economy of scale for process equipment common to all engineering designs. The economics presented in the brief should not claim too much accuracy, because the design presented is usually conceptual and will need further engineering, development, and pilot testing before true economic factors can be estimated accurately.[6]

Report detail. Is the detail provided in each process improvement option sufficient for management to read and form a reasonable conclusion consistent with those in the project team? If not, either more or less detail is needed. Managers should be given *exactly* the same level of detail as in traditional project proposals. Generally, waste reduction process improvements should *not* be presented with too much detail, as there is usually not enough detailed information available. If the pollution prevention effort is at the conceptual stage, use the in-house level of detail for that stage. While this may seem obvious, many engineers tend to provide more detail than is necessary for management approval for implementation.

Incorporation of Improvements into Traditional Design and Implementation Cycle

This is the toughest part of the overall pollution prevention program effort. Here is where all the planning and subtle alignment and support of many areas of the company either pay off or lead to the failure of the proposed process improvement project. Very often, alignment with other process maintenance or planned improvement activities is the way to go.

Incorporation of pollution prevention recommendations with other projects

Pollution prevention must be associated with other, more traditional process changes or improvements. This is analogous to the need to incorporate process control design[7] and energy optimization[8] into the process design work at the earliest possible stages. The sooner pollution prevention improvements can be incorporated into another traditional design project, the better the odds of success.

Often, waste reduction projects lack the economic return on capital investment to justify their installation alone. Many times, they are implemented to comply with current regulations regarding waste generation and release. Further, a number of firms have been required to implement pollution prevention process improvements as part of Environmental Protection Agency (EPA) actions from violations of the Resource Conservation and Recovery Act (RCRA) and other regulations dealing with management and reporting of hazardous wastes. Whatever the reasons for waste reduction process improvements, other process changes and more strategic plans need to be considered before pollution prevention installations are begun.

Optimum program tactics should include consideration of other projects, such as

- Debottlenecking
- Revamps
- Equipment replacements
- Corrective maintenance
- Other capital improvements to the plant

This can result in improved overall economics, less pressure to justify incremental pollution prevention process change economics, lower cost to implement the waste reduction portion of the job, and a better opportunity to tout the future benefits of waste reduction.

Pollution prevention work must be integrated via concurrent scheduling with other process and/or plant changes or improvements. These should include

- Unscheduled start-ups and shutdowns
- Scheduled start-ups and shutdowns
- Product or feedstock changeovers

Further, they should be associated with other existing corporate and site-specific programs for process, product, and project improvements such as health and safety programs, total quality management (see Chap. 1), and certification programs such as ISO-9001 certification. For corporations with research and development operations, long-term waste reduction efforts must be associated with existing technology development programs. This allows pollution prevention to be incorporated into the technology as development occurs. Here, there is little of the opposition evident in those dealing with mature production and manufacturing technologies and processes.

Traditional engineering projects

Pollution prevention incorporation into the process design work for traditional grass-roots or revamp projects is a *very* tough sell. Success can be achieved only if pollution prevention is included and supported in the *proposal stage* by project execution people and project management. This support must involve (1) corporate management, (2) marketing and sales, and (3) project management. These three groups must perceive its value to the project success. If not, pollution prevention will be spoken of at a minimum and will not be considered. The possible reasons vocalized for why it "is not the best thing to do for *this* project" are legion. Moreover, the vocal and usually convincing opposition to including pollution prevention considerations at this very late date will be significant. This is especially true for those who are not familiar with current environmental issues.

Constraints on Effective Implementation

Effective implementation of pollution prevention process improvements has to run through several constraint "hoops." Some of these have been mentioned already. This section attempts to deal with the limitations or constraints involved in any pollution prevention program. They come up at different times, with different people involved, and often are unique to the politics of each organization.

Project demands

Pollution prevention improvements are subject to many demands in project execution. As discussed before, the work must be supported by engineering and project managers, who should be made aware, in advance, of the possibility of incorporation of pollution prevention improvements with the projects they are likely to manage. It will be required that the additional effort not slow down the work process. If the pollution prevention program is run with the same type of management as traditional projects are, and if the pollution prevention options are at the same stage of design as the rest of the process, then incorporation into the existing work process will be possible.

Also, the pollution prevention work, as any project, must not portray any additional risk not already reviewed and accepted. Again, this is why a previously completed program is needed to allow pollution prevention to meld with traditional projects. At this point, any waste reduction addition must have very attractive incremental economics based on verifiable and accepted bases. Also, the technology involved or the changes included must be at the same level of confidence as the rest of the project. Ideally only field-proven, commercially available technology or techniques should be considered. If this is not possible, pilot testing of waste reduction design changes involving new technology or closed loops is a very good idea.

Risk constraints

It is well proved that any process modification becomes a higher-risk candidate as it approaches the core of the production process or technology.[9] This is depicted in Fig. 19.5, which presents the "onion diagram" often used to portray the hierarchy evident in production processes.[10] The risk of any process change is, of course, that the change will disrupt the process with either loss of product or, more importantly, an upset condition that could lead to a safety problem or an accident. The closer you get to the critical production process components, the more confidence is needed in the new design or design change (see Fig. 19.6). A higher certainty of success is required for

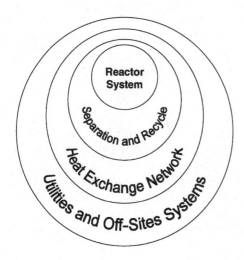

Figure 19.5 Hierarchy intrinsic to chemical processes.

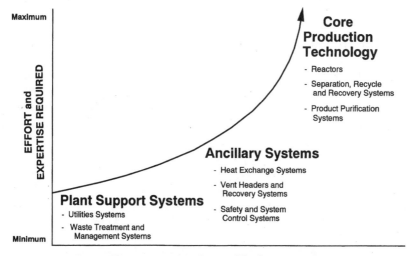

Figure 19.6   Effort and expertise required to modify plant processes.

implementation, and a greater knowledge of the production process is required than in traditional projects. This is because waste reduction designs bring previously discarded streams back into the process, creating a more complex process. Greater complexity usually leads to greater risk. This is why so many experts are needed at the initiation of the pollution prevention project. If credible, respected expertise is involved in the program, it greatly enhances the probability that the pollution prevention alternatives proposed will be accepted.

Time, business, and marketing constraints

Additional constraints involve what groups are available to do the work and the frame of reference of those doing the work. Here, the process design engineers involved must believe that the effort is important to the organization. Again, corporate management support for pollution prevention is important and is manifested in the enthusiasm, or the lack thereof, in the engineers working on the project.

The manager of the program must consider the availability of financial support for large-scale process changes within the company. Often, the timing of the recommendation to implement the pollution prevention improvement is important. Also, the risk aversion of the organization and how it requires addressing the potential consequences of failure of the recommended improvement options need to be considered up front. The manager of the program must know the limitations of the program and why the limitations exist. What is more important, he or she must understand the consequences of exceeding those limitations.

Has the program considered concurrent projects that could erase, alter, or enhance the economics and attractiveness of the potential improvements? Does the structure of the pollution prevention program consider the ultimate marketing aspects of the process improvement? If not, its chances of receiving the required support are slim. This may require discussions with marketing people to gain enough insight to properly format design improvements. Also, does the structure of the program consider the strategic plans of the end user of the process improvement? This is important if the end user is a third-party company or consumer.

How good must the incremental economics be for predictable acceptance in the company doing the work? A "ballpark" knowledge of this will be invaluable early on. What about the market stability of the product produced from the process being improved? Will a slack demand or increased competition reduce acceptable incremental margins required for acceptance? Use past examples and discuss this with those in your company who understand possible trends. If it cannot be determined for any number of reasons, you must keep this in mind when it comes time to sell the idea of a process improvement to management. If the market trends have resulted in poor product margins and a poor cash flow position for the company, suggesting a capital improvement will be poorly received.

Technology constraints

One of the most important constraints is the need for new technology or new applications of existing technology before adequate pilot testing has been completed. This testing is done to reduce the uncertainty and risk involved with any design change—not just for pollution prevention process design improvements. Often, a design to reduce waste generation can lead to adverse production technology performance concerns if consideration has not been given to issues such as safety, performance during upsets, maintenance, and potential changes to plant feedstocks.

During evaluation of new waste reduction designs, facilities must be identified to test the process improvement prior to its anticipated need. Laboratory testing capability and budget must be identified. The testing of the process change must answer the questions raised in the initial option evaluation. It should also clear up or confirm the listed disadvantages identified in the pollution prevention options written discussion. It will also help solidify the advantages stated for the process option.

Dependence of a process option on a single emerging technology where one or two vendors exist is usually considered a concern, for the success of the installation is now dependent on a smaller company that is likely more unstable and less financially secure. This situation should be considered in any ranking methodology used to identify the best options for management consideration and funding.

Information constraints and data availability

The amount of information needed during initial waste reduction design options evaluation is not as great as some would like to think. Often, too much information is gathered and analysis done without thorough consideration of the purpose of the data. The experience and insights of the experts generally present the best picture of the process and the effects of changes being considered. No amount of laboratory testing will provide this. Process operations experts (i.e., the plant operators and maintenance staff) also enhance credibility, as they are generally considered experts by others in the program and by the management who will ultimately approve the recommendations. However, there is one caution. The operations staff need to know why the information is needed and how accurate it must be, before giving out data resident in their heads based on their many years of experience. If this is not truthfully conveyed, it will tell them two things—you do not know why you want the data (i.e., useless data) and you have not conveyed the goals of the program to them.

Additional data sources are also required, of course. As previously noted, there is a great mass of pollution prevention case study data and related information available, and new tools are being developed to improve accessibility to design engineers.

Verification and Sustainable Development

Every pollution prevention program should consider ways to measure the parameters listed below at the onset of each program. By verifying that the results of pollution prevention are as predicted, future pollution prevention programs will be far more likely to receive management support. Recommended measurement considerations concern

- Amount of waste reduction
- Realized operating cost reductions
- Economics of previous implementation projects
- Installation problems that arose and their solutions
- Unexpected changes that occurred in the production process and their effects on the waste reduction process improvement
- Effects of the improvements in waste generation on the environmental permitting time and difficulty for the facility and company as a whole

■ Costs of end-of-pipe treatment and waste management before and after implementation of the waste reduction improvement

Verification enables future pollution prevention efforts to be more easily implemented and budgets to be increased. It also improves good relations and enhances respect for the overall benefits of waste reduction programs.

Establishment of new goals

To maintain the momentum of the pollution prevention program, realistic new goals must be established. These new goals and the measured successes in the past will lead to more available resources to continue the effort and will improve the support of upper management. This can be particularly important when pressures mount to cut overhead, long-term programs, and anything that does not produce a direct benefit to the bottom line.

Repeating pollution prevention methodology

All pollution prevention programs should stress that sustainable reduction of waste generation can occur only with a continuous process of improvement. Pollution prevention programs are *not* once-through efforts. If they are deemed so, they will disappear quickly. Moreover, many goals set by these programs can be achieved only by incremental improvement based on alignment with other traditional projects, as mentioned previously. More importantly, environmental regulations are constantly changing, and they are becoming far more restrictive than many believed possible. As additional regulations are promulgated, new technologies will be developed to address the new demands. Only through a continuous program will these changes be effectively conveyed to the organization. Also, no design is so good that it cannot be improved. New people with additional experiences will join the organization and will bring with them new insights and abilities. Remember, new people in new situations with new teams will produce new process designs. It has been this way for many years for all types of engineering. Why should pollution prevention be any different?

References

1. Stephen W. Morgan, "Use Process Integration to Improve Process Designs and the Design Process," *Chem. Eng. Prog.*, September 1992, p. 67.
2. Agnes Stanley, "Pollution Prevention: Reinventing Compliance," *Chem. Eng.*, November 1993, p. 31.
3. D. W. Hertz, P. P. Radecki, B. A. Barna, T. N. Rogers, D. W. Hand, D. R. Shonnard, A. S. Mayer, M. E. Mullins, J. C. Crittenden, D. F. Rudd, D. C. Luehrs, J. R. Mihelcic,

and D. A. Crowl, "Development of CPAS: A Pollution Prevention Conceptual Process Design System," Presented at the 207th American Chemical Society national meeting, San Diego, March 12, 1994.

4. Kimberly A. Roy, "Attitude Adjustment," *Hazmat World,* May 1989, p. 25.

5. Agnes Stanley, op. cit., p. 39.

6. Carl H. Fromm, Michael S. Callahan, Harry M. Freeman, and Marvin Drabkin, "Succeeding at Waste Minimization," *Chem. Eng.,* September 14, 1987, p. 94.

7. Robert E. Sheffield, "Integrate Process and Control System Design," *Chem. Eng. Prog.,* October 1992, p. 31.

8. Morgan, op. cit., p. 62.

9. Elizabeth A. Flores, "Impediments to Haz Waste Minimization," *Pollution Eng.,* March 1991, p. 78.

10. Robin Smith and Eric Petela, "Waste Minimisation in the Process Industries," *The Chem. Eng.,* October 31, 1991, p. 25.

Chapter

20

Standardized Accounting for a Formal Environmental Management and Auditing System

Mohamed A. Serageldin, Ph.D.
Triangle Speakers Group
Chapel Hill, North Carolina

Abstract

Obtaining reliable data is one of the major problems facing managers engaged in pollution prevention activities. The quantification of process emissions is notoriously difficult; and the problem is made more complex by the many inconsistencies that arise in the measurement and reporting of process releases. A standardized methodology of accounting, enabling the manager to make meaningful comparisons between reported performances of different processes, plants, or equipment types, is an essential requirement for good decision making. This chapter describes such a methodology.

The procedure is based on the *unit operation system* (UOS) concept. This prescribes a systematic and unified approach for developing a materials-tracking inventory that can satisfy the needs of plantwide programs for evaluating pollution, cost, and risk. It provides the ideal approach for developing a general inventory to store emission data, materials use data, and waste data. It can be used to assess the cost-effectiveness of pollution prevention or pollution reduction options, and it helps companies to identify those sections of their processes that are most problematic in terms of emissions. This, in turn, helps to focus management and engineering effort in the most fruitful areas. At the national level, identifying problematic UOSs can enhance cooperation between research organizations, companies, and vendors, to develop new or innovative strategies with a focus on pollution prevention and novel control technologies. The UOS approach also provides natural building blocks for performing *life-cycle analysis* (LCA) and for making decisions regarding rebuilding or downsizing.

Introduction

It makes economic sense to adopt a standardized environmental accounting system that promotes more defensible decisions and enhances communication between corporate management and the plants being monitored within an organization. A standardized accounting procedure enhances data accuracy and leads to a more systematic procedure for data analysis and quantification of the pollution generated by a plant. Moreover, when many plants in the same industry adopt the same accounting methodology, there is a common basis for making comparisons of similar categories of pollution sources.

This chapter presents aspects of sound accounting, using a novel concept, to obtain data of *high specificity,* to evaluate alternatives, and to perform plant audits. The concept was developed by this author a number of years ago and was named the *unit operation system* (UOS). UOSs are the building blocks on which the concept is founded. Data presented according to the indicated UOS format facilitate the making of management decisions aimed at pollution prevention and reduction. It also identifies essential information that needs to be reported and monitored, thus minimizing unnecessary pollution-related record keeping.

This discussion will focus on solvent cleaning in industry. In this way we will deal with only one type of material. This will facilitate the discussion of the UOS concept.

Data Quality and Accuracy

Plant-specific pollution information is commonly acquired by means of surveys (or information requests). Such information is used by industry, trade organizations, and environmental groups for different purposes. The quality of the data, however, depends on both the *specificity* of the questions asked and the *quality* of the responses received from the plant surveyed. Compiling data in an efficient and consistent way across a given industry is usually a major problem, especially in the case of *fugitive* pollution, such as that from solvent cleaning. The same can be said of *secondary* pollution, which results from recycling process waste or storage of the waste within the plant boundary. It is understandable that the data are not consistent, given that accounting procedures used by different plants, even in one industry, are unlikely to be the same. The required data may not be readily available, and assumptions have to be made to generate the requested data.

This problem becomes much more visible when a large plant population is being surveyed. The task of surveying is made more difficult by the nonuniformities and possible inconsistencies in the responses received. The issues raised here are important because they influence the level of accuracy of the data collected and thus the value of the quantified pollution released in air, water, and land. Adopting the UOS concept across an industry offers a viable and economical solution. Equivalency can be evaluated accurately and realistically at this level.

Inventory Assessment (Accounting by UOS)

Let us assume that we are a trade organization and that our purpose is to survey plants in the industry to identify pollution prevention and pollution control practices, to develop options for the industry. The survey should be designed so that the information received will lead to a determination of the "ideal" block (group) of pollution points or block of equipment items, on which the alternatives are to be evaluated, i.e., the unit operation systems.

The boundaries of each unit operation system within a plant should be drawn so that *accurate* and *specific* assessments of the components of pollution in air, water, and land can be made. Pollution limits and *equivalency* can later be based on the UOSs. The definition of a UOS should be sufficiently specific for a plant to be subdivided (to account for pollution) into categories of UOS that are common to the industry.

Total plant use of a material, volatile emissions, and waste can be determined from the sum of the corresponding values assigned to the different UOSs. For example, if the purpose of the inventory is to determine evaporative emissions from plantwide solvent use, we should be able to obtain this value by adding values entered for each UOS, as indicated in Table 20.1.

Each UOS emission entry is obtained as follows:

$$\text{Solvent emissions} = \text{input solvent} - \text{output solvent} \quad (20.1)$$

$$\text{(UOS)} \qquad\qquad \text{(contaminant-free)}$$

Total solvent use based on the UOS should be checked against total plant solvent use (and not from general inventory values). Also the mass of sol-

TABLE 20.1   **Summary Table of Solvent Emissions from UOSs**

Solvent type	Solvent emissions (Mg/yr)						Total
	Unit operation system						
	1	2	3	4	5	...10	
A	17.0	4.0	6.0				27.0
B				14.0			14.0
UOS total	17.0	4.0	6.0	14.0			Plant total
							41.0

Note: Solvents *A* and *B* are 100 percent VOCs.

vent waste (free of contaminants) from the UOSs should be checked against the total mass of solvent waste removed from the plant boundary.

If either the use values or waste solvent values do not check, plant managers should introduce the necessary corrections to account for the difference. The procedure used for making the corrections for material balance closure should be recorded. The difference between the total plant value and the value based on the sum of the UOS values provides a measure of the level of plant managers' *materials awareness* (MA). An acceptable level or perhaps a range of values could be determined to define management's MA on a scale of low, medium, or high.

Possible Levels for Accounting

To quantify air emissions by using the UOS concept, we need to identify the emission points and the pollutants from the different UOSs. Before taking this step, we need to identify the areas in the plant where key production activities are occurring, so that we can identify the possible sources of pollution.

There are a number of possible levels[1] at which pollution accounting can be based:

- A *plant* consists of one or more departments.
- A *department* consists of one or more work areas.
- A *work area* consists of one or more unit operation systems.
- A *unit operation system* consists of one or more unit operations.
- A *unit operation* consists of one or more *items of equipment,* which may be cleaned by one or more *cleaning activities,* such as wiping and flushing.

Figure 20.1 provides a visual representation of the position of a UOS in a department. However, a plant may have the same category of UOS

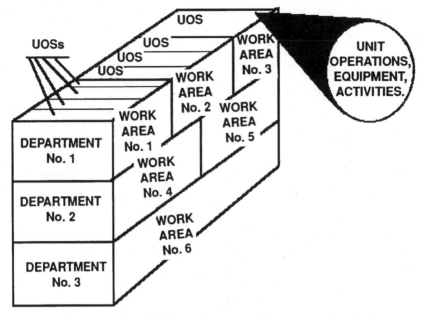

Figure 20.1   Representation of UOSs, work areas, and departments in a plant.

(e.g., spray booth) in more than one department or work area; and there are instances where only one UOS is defined in a work area, such as in Fig. 20.1. When the accounting is performed above the UOS level, one cannot distinguish between the different pollution sources in a department, and thus one cannot properly evaluate equivalency. That is why it is recommended to evaluate pollution at the UOS level (or at the unit operation level) rather than at the plant or other nonspecific (pollution) level. A cost (or business) center level, which normally includes one or more departments and several work areas, is a nonspecific level, as can be seen from Fig. 20.1.

What Is a Unit Operation System?

The term *UOS* refers to a formalized concept for performing a material balance. A *unit operation system* means the ensemble on which the material balance is performed. If the purpose of the material balance is to determine emissions due to the use of cleaning solvents, it encompasses all possible points and sources leading to evaporative emission losses associated with the cleaning of a primary *unit operation,* including losses during dispensing of solvent, handling of residuals in cleaning tools (such as rags), solvent storage, and so on. Figures 20.2 and 20.3 are examples of *primary* UOSs that are commonly used with cleaning solvents and for the painting of exterior surfaces. These two

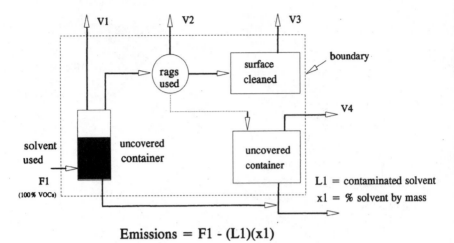

$$\text{Emissions} = \text{F1} - (\text{L1})(\text{x1})$$

Figure 20.2 Representation of a primary UOS for cleaning surfaces: uncontrolled emissions, uncovered containers. (Source: Ref. 1.)

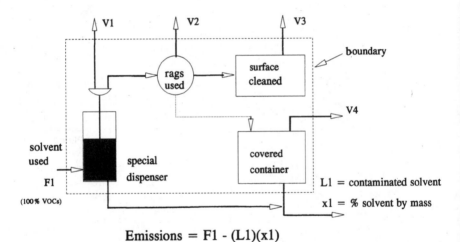

$$\text{Emissions} = \text{F1} - (\text{L1})(\text{x1})$$

Figure 20.3 Representation of a primary UOS for cleaning surfaces: controlled emissions, covered containers. (Source: Ref. 1.)

figures illustrate different pollution control strategies, which are discussed in more detail later in this chapter.

If the purpose of the material balance is to determine emissions from a chemical process which involves optimization of the performance of a reactor attached to a distillation unit operation (the primary unit operation), it includes all the items of equipment that are *physically integrated* by means of piping or conduits. *Secondary* pollution as a result of storage of raw material or storage of waste should be added to the emissions from the primary unit operation.

A *unit operation* means an industrial operation, classified or grouped according to its function in an operating environment. The definition for our purpose includes a paint-mixing vessel (tank) and ancillary piping and equipment, a spray booth, a parts cleaner, and a printing machine, in addition to a conventional separation unit operation such as a distillation column with an attached condenser and an attached reboiler. A unit operation (UO) may, therefore, involve one or more items of equipment; e.g., a unit operation may include several mixing vessels, both a reactor and a mixing vessel, or a reactor and a distillation column. More complete definitions of UOS and UO are given in Ref. 1.

Unit operation systems (and unit operations) may conveniently be categorized into primary and secondary UOSs (and UOs). The primary unit operations and UOSs represent the production units, such as mixing tanks and attachments needed for manufacturing paint. Associated with this process is secondary pollution resulting from storage of raw material such as solvents and waste solvent. Secondary pollution also results from steps taken to recycle and reclaim chemical waste. The emissions from a UOS are, therefore, the sum of the corresponding primary and secondary UOSs. This statement is illustrated in Fig. 20.4, which shows the main components of a UOS. Ideally any transportation losses to and from the UOS should also be apportioned by the UOS.

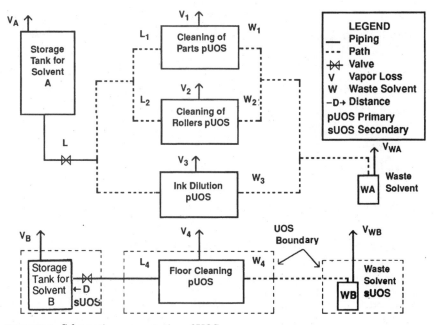

Figure 20.4 Schematic representation of UOSs.

Some Formal Developmental Steps

When the UOS methodology is applied to a number of plants which manufacture paint or to plants involved in surface coating, we are able to identify several generic categories of UOSs. Four are common to manufacturing, painting, and cleaning: (1) the spray booth, (2) the spray gun, (3) the tank, and (4) the line. A *line UOS* will include, as in the case of a tank UOS, the mixing tanks (process vessels) and all the lines physically connecting the different items of equipment through which process material such as paint flows. The word *line* or *tank* is used to indicate whether most of the surface area in contact with the process material is due to the connecting lines (piping) or to the tanks. Both areas have to be determined to normalize emissions or waste from a UOS, as will be discussed later. The line unit operation is found, e.g., in plants where painting is done in spray booths located 30 to 40 m from where the mixing of the paint occurs.

Before a manufactured part, such as the body of a car (or a small metal box), is painted, its surface area has to be cleaned. This is often done with cleaning solvent. An item of equipment such as a tabletop or a pump is often cleaned with solvent. Removable parts are cleaned in a parts cleaner. The floor in a plant may require regular cleaning. If the paint manufactured or applied is solvent-based, then a solvent-based cleaning agent must be used to remove the paint from the floor, unless the floor is attacked by the solvent. A UOS can be developed for each of these cleaning UOs to determine the pollution generated. Illustrations similar to Figs. 20.2 and 20.3 can be used to represent UOS categories involving painting of exterior surfaces or cleaning of parts.

The following steps are required to develop the data for a UOS:

1. Identify and briefly describe the primary unit operation (UO).

2. Draw the boundary of the UOS.

3. Identify all input/output streams and emission sources crossing the UOS boundaries.

4. Indicate the level of control (e.g., open container, special solvent dispenser, or special cleaning tools).

5. Identify the waste streams, and determine the amounts of solvent and contaminant in the waste streams.

6. Indicate the boundaries of a UOS closed-loop (in-house recycling) units.

7. Identify fugitive units such as valves.

8. Identify the characteristic dimension (e.g., surface area painted, surface area cleaned, evaporative surface area, material input rate, product output rate, or time elapsed) for that UOS.

When the material tracked is an organic solvent, the number of secondary UOSs for the above example depends on the manner in which the virgin solvent is stored and brought to the primary UOS and the means by which the waste is handled beyond the boundary of the primary UOS. If the virgin solvent is stored in large tanks, the plant will need to estimate the amount of evaporative emissions from these dedicated solvent storage tanks. An approved analytical method is used for each solvent, and then the estimated losses are apportioned by primary UOS. Sometimes the solvent is recycled in house or is stored prior to removal by a waste-handling agency. In such cases, the plant needs to develop an independent UOS for each in-house recycling unit operation, reclamation unit operation, or waste solvent storage area. It will then need to apportion the emissions by primary UOS. These steps are necessary to evaluate equivalency properly.

The Material Balance

The boundary of a UOS has to be explicitly defined, as shown in Figs. 20.2 and 20.3; otherwise, a meaningful material balance cannot be done.* Different measures are taken to reduce solvent pollution in the two cases illustrated here. In Fig. 20.2, the solvent container and the container for soiled rags have no covers, so evaporation occurs freely. In Fig. 20.3, we show a solvent dispenser being used, and the container for the soiled rags has a tight lid. We refer to the second case as a *controlled* UOS, whereas Fig. 20.2 represents an *uncontrolled* UOS. The practices used in Fig. 20.3 are pollution prevention practices, since they are aimed at reducing pollution at the source. The source in this case is defined by the UOS boundary (and not the boundary of the primary UOS shown in Fig. 20.2 or 20.3).

When the boundary of each UOS has been properly outlined, information at the UOS level provides the appropriate level for defining baseline pollution. It also provides a basis for defining pollution limits and for monitoring progress in pollution reduction and material waste minimization. All input streams and output streams crossing the UOS boundary must be identified on the drawing, which becomes part of the plant records. Such drawings are key to the UOS methodology.

If a plant wishes to monitor pollution correctly at the UOS level, or if it wishes to recommend pollution prevention practices, first the plant must ensure that the data collected are both accurate and specific. The

---

*These figures depict only primary UOSs, because the solvent storage and waste tanks are not included in the drawing. In the completed UOS, the virgin and waste solvent storage tanks would have been clearly labeled in the drawing; otherwise, the emission data and waste data could be misunderstood.

difference between an overall plant material balance and one based on the sum of the UOSs provides a cross-check needed for determining accuracy. The difference can be determined as follows:

$$\text{Emission difference} = L - W_L - \left(\sum L_N - \sum W_N \right) \qquad (20.2)$$

where L is the total use of solvent, W_L is the total waste solvent (free of contaminants), and L_N and W_N refer to actual input and output solvent streams, respectively, for N unit operation systems. In Fig. 20.4, N takes values from 1 through 4. We should include in Eq. (20.2) the losses from storage of virgin solvent and from storage of waste solvent.

A plant should not, therefore, operate without defining the acceptable level of uncertainty. This point will be taken up in the next section.

A Case Study

We will deal with an imaginary plant whose main business is producing flexible packaging for consumer and pharmaceutical uses. The plant houses a number of rotogravure and flexographic printing presses, extruders, and waxers. The machines are operated intermittently depending on customer demand. In this example, we assume that the plant used only two types of solvents. One solvent (solvent A) is used for both cleaning and diluting the ink (manufacturing), and the other (solvent B) is used for cleaning only. This plant uses solvents to clean rollers and print cylinders on the printing presses—parts such as ink pans and pump parts; to clean tables; and to remove contaminants from the floor.

A matrix of solvent versus UOS (Table 20.1) was developed by the plant, to summarize solvent emissions from all UOSs, both primary and secondary. The overall plant emissions given at the bottom of the table were calculated by adding the emissions from the tabulated UOSs. (Similar tables can be developed for solvent use and mass of waste solvent.)

A comparison between the sum of emissions from all the UOSs and the emissions from a plantwide material balance on solvent cleaning showed a discrepancy of 20 percent (mass basis). This warranted revision of the assumptions made and reevaluation of the maintenance records. The revision process can be done more effectively when the plant records include simplified schematics illustrating the manner in which the solvent is brought to, and waste solvent removed from, each UOS (for example, Fig. 20.4).

Solvent A was carried in small open-top containers to each task. It was used as needed by turning on a valve attached to the main storage tank. The plant did not keep separate records, which would have enabled it to determine the amount of solvent A used for cleaning the rollers and parts and the amount used for diluting the ink. The supervisors had to estimate the solvent distribution between those two distinct tasks, after consulting with the plant employees. Another assumption was made to determine the input streams to the UOS for cleaning the parts and for cleaning the rollers on the printing presses. The plant manager realized that the plant had not reported any solvent used for cleaning the tables. The solvent used here was estimated to be about 10 percent of that used cleaning the parts. Solvent B was used only for floor cleaning, and the waste from that cleaning activity was stored separately from the waste generated from solvent A. The broken lines in Fig. 20.4 indicate that the transfer of solvent was performed manually.

The plant did not know, by UOS, how much contaminant was in the solvent waste from cleaning. The exception was floor cleaning, where the contaminant was estimated to be around 15 percent by mass. Waste solvent and used rags from cleaning were transferred in small containers, located at various activity locations, into 55-gal drums. These were stored in a general waste collection area. The plant had estimated that the average contaminants in solvent A (excluding the mass of the rags) represented around 30 percent by mass. (The contaminant in waste from solvent cleaning varies from one industry to another and can exceed 40 percent by mass.) In view of the many assumptions made by the plant personnel, it is not surprising that the emissions by UOS fell short by the stated 20 percent.

Simplified flow diagrams like Fig. 20.4 provide a useful tool for discussion and for instruction to plant personnel. One can easily deduce from such a figure that solvent use is not adequately regulated or monitored. The addition of metering devices may not be necessary in this situation. Assigning a solvent storage tank for each task would increase accuracy. The use of special dispensers as shown in Fig. 20.3 (controlled situation) will reduce spillage and the use of solvent (pollution prevention!). Time and money spent redoing the calculations would also have been saved. A plant which adopts a pollution prevention focus needs to operate within a well-defined overall uncertainty level.

Pollution Data and Secondary Users

The example discussed in the previous section is not intended to be an exhaustive exposition of the concept and underlying methodology. Rather, it highlights a number of factors that can help increase confidence in the data being generated. This is important, given that emission data may be stored in a database for later use. If the methodology is used correctly, secondary users will have an understanding of the nature of the assumptions made during the quantification stages.

There will also be some guarantee that the entries in a database for each source were generated in a consistent way. As indicated above, uncertainty in emission (or waste) data determination is a problem. The problem is more manageable if one can quantify the *error range*. Plant management would have to specify the acceptable variance (tolerance), by UOS category (or subcategory, if necessary). The challenge would be to define an acceptable level of variance. This is an important issue which is relevant to verification of a plant's compliance with management directives. When UOS data across an industry become available, an error range can be readily determined.

Unless we know the magnitude of the error, we cannot assume that the values for a type of source in a database belong to the same statistical population. Consequently, using such a database to generate pollution factors for pollution determination could result in inappropriate decisions.

Normalizing Emissions from UOS and Waste Streams

Use, emissions, or waste stream data may need to be normalized for a number of purposes, including comparisons between the performances of different plants, evaluation of pollution prevention options and permitted pollution levels, and determination of emission factors. The normalizing parameter selected should represent the important factors that influence the emission rate. In the case of solvent cleaning, e.g., the surface area cleaned, hours used for cleaning, and solvent used for cleaning invariably influence air emissions. In practice, one should strive to develop emission factors by using only one parameter, e.g.,

$$\text{Parameter} = \text{surface area} \qquad (20.3)$$

When the surface area cleaned X_1 and time for cleaning X_2 are both important parameters, then

$$\text{Parameter} = f(X_1, X_2) \qquad (20.4)$$

Figure 20.5 shows a plot of mass of evaporative emissions for a number of UOSs versus surface area cleaned. Such data points would be provided by different plants in one industry. The slope of each solid line represents an average normalized emission (an emission or pollution factor). When the data fall around two distinct lines, we can deduce that the data belong to two different populations of the same category of UOS (e.g., a new technology and an old one).

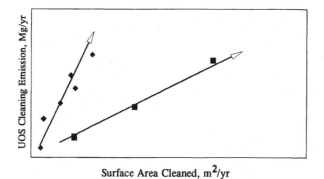

Surface Area Cleaned, m^2/yr

Figure 20.5   Plot of emissions versus surface area cleaned.

When the slopes are determined by linear regression, we can also determine the variance attached to each value. Plots similar to Fig. 20.5 can be generated to determine normalizing parameters for manufacturing UOSs. The applicability of this concept to manufacturing operations is discussed in Ref. 2.

When UOS pollution factors exist for all categories (and subcategories) of UOSs in an industry, they can be used to set plantwide achievable environmental goals. UOS pollution factors can also be used to determine acceptable release levels for the same category of UOS. This concept is important for rebuilding or downsizing. Pollution factors derived in this way aid in determining options for situations where there is variability in the numbers and sizes of UOs in a work area. The variability may also be attributable to the different activities occurring within the plant.

Record Keeping

A plant keeps records to perform internal audits and to monitor activities that reduce or prevent pollution. Both activities can be standardized if the companies in an industry agree to adopt the UOS concept for data collection, pollution accounting, and equivalency determination. This makes sense as more companies work closely with their respective trade organizations as a means of sharing information. What are the additional benefits of adopting a standardized accounting system? I can think of several: a reduction in the cost of accounting; shorter turnaround time for plant responses to information requests; higher accuracy and specificity of reported data and of pollution quantification; and more defensible management decisions and actions. Management can use the highly specific information obtained in this way to set standards of work performance, UOS emission limits, waste limits, and plantwide limits. The data requirements for monitoring performance are also simplified. Furthermore, the UOS approach helps to identify core pollutant unit operations, so that a focused management effort will lead to large improvements in air quality or a significant reduction of emissions to other media.

The UOS information also can be used by a plant to develop inventories of pollutants and to evaluate residual risk. The data can be conveniently updated. The visual representation required for each UOS is an essential requirement of this methodology.

Data entry for a UOS can be done manually or automatically on a computer. It is surprising that not too many industrial companies are taking advantage of hand scanners (such as those used by small and large grocery stores) to enter plantwide data from central material disbursement locations. Solvent (and other material) use, worker-hours expended, and pollution generated can then be properly assigned by unit operation system. Thus automation can effectively reduce the cost of record keeping and increase the accuracy of the data that are recorded.

There is an additional benefit of maintaining records at the UOS level. Emissions or wastes generated from a UOS may increase or even decrease even though the plantwide unit of production remains constant. By assigning material use by UOS, a plant will be able to quantify the effect that a change in plant operations has on pollution. This point is especially important if the plant is trying to determine the cause of the change in pollution and whether a true reduction in pollution has occurred.

Cost-effectiveness

The *cost-effectiveness* (CE) of an option for reducing pollution is an important parameter which can influence the outcome of a decision. It can be defined as follows (Mg = megagrams):

$$CE = \frac{\text{net increase in cost (\$/Mg)}}{\text{evaporative emission reduction (Mg/yr)}} \qquad (20.5)$$

Adopting the UOS concept would require that the solvent be tracked from cradle to grave for each UOS in all three media (air, water, and land). The VOC emission reduction would include the sum of the reductions for both the primary and all secondary UOSs shown in Fig. 20.4. A plant may be required to adopt pollution prevention practices and to maintain records. The costs associated with these activities should be included in Eq. (20.5). When add-on controls are required to offset cross-media pollution, their cost must also be counted. Cost-effectiveness is not, therefore, based on considerations of a single medium or of an isolated item of equipment.

Discussion and Conclusion

A material balance can be performed at any of the plant levels outlined above. However, the UOS is the ideal level of accounting for the purpose of managing pollution (e.g., pollution prevention), for assigning the costs of pollution, for determining the suitability and cost-effectiveness of options, and for assessing risk. By adopting this focus, one is able to develop a "generic" accounting system that is founded on modules which accurately reflect all the pollution and costs at the source of pollution.

The UOS concept helps to divide a plant into a matrix of categories that are common not only to a given industry but also to other industries. Consequently, the UOS approach provides a consistent basis of evaluation that can be used across an industry, and even between industry sectors, irrespective of the organizational or process differences at different sites. This is a major advantage. *Cost centers, departments,* and other customary groupings do not offer the same consistency, and they are therefore unsuitable for comparing emission rates or evaluating the benefits of pollution prevention or pollution control options. The company must address pollution issues at the unit operation system level.

The UOS approach provides natural building blocks for performing a life-cycle analysis. It also provides a useful tool for tracking improvements in air quality, since the elements for verifying that the limits are being met are available.

Another major application of UOS accounting is the identification of the types of process operations that are most likely to be major sources of emissions. It highlights the benefits of alternative technologies that reduce pollution. This helps to focus management and engineering effort where it is most needed and offers a number of benefits, of which perhaps the most important are these:

- Identification of "problem" UOSs helps to focus research, leading to the development of novel strategies for solving the underlying pollution problems.

- Emission data and pollution prevention solutions can be transferred more readily between production facilities (see also Chap. 11).

- The quality of management decisions relating to pollution prevention and pollution control is improved.

References

1. M. A. Serageldin, "An Illustrated Glossary of Terms and Definitions with a Focus on the Unit Operation System," No. PB-95-144655, National Technical Information Service, Springfield, Va., July 1994, pp. 1–31.
2. M. A. Serageldin, "Multi-Media Accounting and Management: Applicability of the Unit Operation System Concept to the Chemical Industry," American Institute of Chemical Engineers, 1994 spring national meeting, paper 36d, Atlanta, April 17–21, 1994.

21

Overcoming Decades of Distrust: The Amoco/EPA Yorktown Experiment*

Ronald E. Schmitt
Amoco Corporation
Chicago, Illinois

Howard Klee, Jr., Sc.D.
Director of Environmental Affairs and Safety
Amoco Corporation
Chicago, Illinois

Debora M. Sparks
Amoco Corporation
Cooper River, South Carolina

Mahesh K. Podar, Ph.D.
U.S. Environmental Protection Agency
Washington, D.C.

In late 1989, Amoco Corporation and the U.S. Environmental Protection Agency (EPA) undertook a voluntary project to study pollution prevention opportunities at an industrial facility. The project grew out of ongoing discussions between the agency and Amoco regarding regulatory

---

*This chapter is adapted from "Overcoming Decades of Distrust: The Amoco/EPA Yorktown Experiment," Corporate Environmental Strategy, 1(2), October 1993. Reprinted with permission of PRI Publishing © 1993.

reforms in such programs as the Resource Conservation and Recovery Act. It followed that a multimedia look at an entire facility would be educational for all involved. The project would also coincide well with ongoing agency efforts to define pollution prevention. Amoco proposed use of its refinery at Yorktown, Virginia, to conduct a multimedia assessment of releases to the environment and to develop and evaluate options to reduce these releases.[1]

Amoco's Yorktown refinery is a 35-year-old, 53,000 bbl/day facility that manufactures gasoline, heating oil, liquefied petroleum gas, sulfur, and coke. It is located on the York River in Virginia, near the Chesapeake Bay and very close to Washington, D.C. By the end of the project, more than 100 state and federal government officials had visited the facility for a first-hand view of refinery operations and practices.

In the end, the project showed that the Yorktown refinery, given more flexibility, could achieve greater environmental protection at one-quarter of the current cost of compliance with existing regulations. This opportunity could mean a saving of $40 million in pollution control costs at one facility alone.

Risks in Forming a Partnership

Developing a partnership required that the EPA and Amoco overcome fears of the risks involved in a cooperative effort. First, there was the unknown expense of the undertaking, an expense that eventually totaled $2.3 million. Amoco funded about 70 percent, and the EPA provided the balance. The project would also require people to work on further research growing out of the Yorktown study itself, work that would be in addition to or in place of their regular jobs. And no one knew whether the time, labor, and money invested would produce worthwhile results.

Amoco also recognized a risk in opening its operations to scrutiny by regulators. It was possible that the study would find an unknown problem or inadvertent violation of regulations at the refinery. The company could be liable for fines or suffer other unwelcome consequences. Amoco believed, however, that it had nothing to hide. The company wanted its operations to be in compliance with regulations, so any information that contributed to compliance was worth the risk.

For its part, the EPA saw a risk in appearing too close to a company that it regulated. Working on a joint project would inspire critics to charge coziness between the agency and Amoco. In addition, a joint project had never been done before, so the Yorktown project was establishing a precedent, and the new partnership would involve the risk associated with any new venture. Acknowledging the risks yet anticipating the benefits, Amoco and the EPA decided to go forward. With no guarantees or promises, the two parties became partners with a simple verbal agreement. Environmental strategists in the 1990s must confront and manage such risks.

Making the Partnership Work

To see the project through, a work group was formed, bringing together representatives from the Commonwealth of Virginia, the federal government, and Amoco. The work group identified four objectives for the study:

1. Define the chemical type, quantity, source, and medium of refinery releases to the environment.

2. Develop options to reduce selected releases, and rank the options on a variety of criteria and perspectives.

3. Identify and evaluate factors that impede or encourage pollution prevention.

4. Enhance participants' knowledge of refinery and regulatory systems.

At the work group's request, the EPA arranged for Resources for the Future, a nonprofit, environmentally oriented policy research organization, to assemble a group of outside scientific and technical experts. This peer-review group provided evaluation and advice on the project work plan, sampling, analysis, results, and conclusions. In all, more than 200 people and 35 organizations with expertise in many disciplines became involved in this project. Their participation reflected the work group's belief that solving difficult environmental problems requires the cooperation of many of society's partners.

A look at Yorktown's process flow diagram gives a sense of potential release sources at this facility (see Fig. 21.1). Each icon in Fig. 21.1 represents an opportunity for pollution prevention strategies. The EPA's pollution prevention program encourages the reduction of pollution generation at the source so that less end-of-pipe pollution management is needed. The work group adopted this general concept but agreed to consider all opportunities—source reduction, recycling, treatment, and environmentally sound disposal—as potential options in pollution management. An environmental sampling program helped identify specific release sources in the refinery that could be targeted for potential process or operational changes to reduce their pollution and environmental impact.

A central goal of this project was to identify criteria and develop a system for ranking environmental management opportunities that recognized a variety of factors, including release reduction, technical feasibility, cost, environmental impact, human health risk, and risk

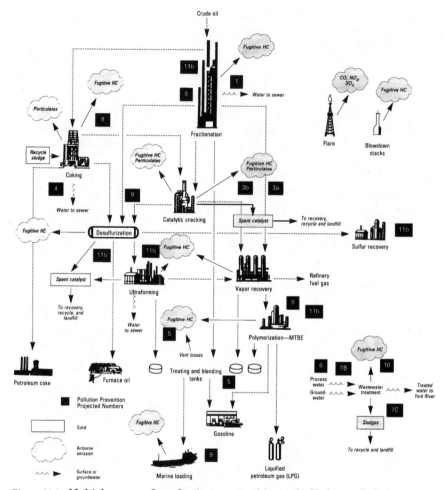

Figure 21.1   Multiple sources for reduction opportunities at the Yorktown Refinery.

reduction potential. Due to the inherent uncertainties in risk assessments, the project focused on relative changes in risk compared with current levels, rather than establishing absolute risk levels. Because of difficulties in quantifying changes in ecological impact from airborne emissions, changes in relative risk were determined based primarily on the human health effects indicated by changes in exposure to benzene (through measured and estimated ambient benzene levels).

The project incorporated many components associated with pollution prevention and facility management choices, including pollution source identification, sampling, exposure modeling, risk assessment, engineering, economics, meteorology, and public perception. The engineering work included the use of process integration techniques, as discussed in Chap. 12. Several areas, however, were specifically excluded from this project. For example, sampling time and data provided a snapshot of releases rather than actual measurements of average annual values, and chemical changes of airborne pollutants were not evaluated. Also, available technologies were considered rather than innovative techniques or those that would require research and development. And finally, emergency and upset events were not studied in this project.

Pollution monitoring for compliance with discharge limitations typically occurs at the point of discharge rather than at the point of generation. As a result, the project needed to collect detailed, chemical-specific release information for use in characterizing the refinery's baseline pollution reduction potential—an expensive and time-consuming undertaking. To do so, the work group developed a sampling and monitoring program that included about 1000 samples. The probable accuracy of most measurements is ±100 tons/yr. Each sample was analyzed for 15 to 20 chemicals, such as benzene, toluene, ethylbenzene, and xylene, as well as for particular chemical species, such as metals and polynuclear aromatics.

The sampling program took about 12 months to complete at a cost of approximately $1 million. Even with this time and dollar commitment, however, only selected sources were sampled. The final release inventory was assembled by using a combination of actual sampling, measurements, dispersion modeling, and estimates based on emission factors. This effort resulted in the first major database showing all releases from a single refining facility into all environmental media at one point in time.

Managing Refinery Releases

This release inventory process enabled a comparison of pollutant generation, on-site management, and ultimate releases to the environment. It showed that the refinery generates about 25,000 t/yr of pollutants [1 metric ton per year (t/yr) = 1.102 (U.S.) tons per year (tons/yr)]. As a result of site hydrogeology, on-site wastewater treatment, and solid waste recycling practices, about 11,000 t/yr is recovered, treated, or recycled and does not leave the refinery site. Of the remaining 14,000 t/yr, about 90 percent is released to the air. Once this fact was recognized, the work group focused much of the remaining project resources on the largest releases—airborne emissions.

Modeling studies done as part of the project indicated relatively little naturally occurring transfer of hydrocarbon emissions from air into other media.[1] Most hydrocarbons are not very water-soluble and so are not easily removed from the air by rainfall. Although the fate of criteria airborne pollutants (for example, NO_x and SO_2) was not studied in this project, they are known to be scavenged by rainfall and can contribute to nitrogen loads and pH changes in lakes and soil. Measurements and modeling results for the project reveal that small transfers from some surface water ponds at the site to groundwater was occurring. Groundwater also enters the wastewater treatment system through the underground sewers, resulting in a net groundwater inflow.

Intentional transfers of pollutants between media commonly occurs as a result of pollution management activities. More than 335 t/yr of hydrocarbons initially present in the refinery's wastewater system volatilizes into the air. And more than 1800 t/yr of biosolids requiring disposal is produced by biotreating wastewater in the refinery's activated sludge system.

Managing Public Perception

The Yorktown project was unique in that the work group wanted to include information about York County residents' perception of the refinery's impact on the surrounding environment and its place in local environmental concerns. Three distinct activities were undertaken to collect this information—interviews with community leaders, meetings with focus groups, and a telephone survey of 200 households.

These efforts found land development to be the major quality-of-life concern in the Yorktown area. People cited traffic, problems with sewer and water services, and general development as the major detractors from the current quality of life—not the refinery. Residents also indicated concern for the reduced fishery yields in the Chesapeake Bay. There has been no clear linkage between the refinery and the yields.

The survey also identified a need for a credible communication channel that the refinery and community could use, if needed. There seemed to be no authoritative source of information about environmental problems that people accepted as reliable.

A Public Workshop: Its Strategic Value

The release inventory and other data collection efforts served as the basis for identifying ways to reduce pollutant releases. More than 120 people from the EPA, Amoco, the Commonwealth of Virginia, and academic, environmental, and consulting organizations participated in a 3-day brainstorming workshop, held in Williamsburg, Virginia, in March 1991. The workshop participants generated more than 50 potential release reduction options for the refinery. These ranged from producing a single grade of gasoline to specific technical options for particular equipment or processes.

The work group subsequently narrowed this list to 12 options for more careful, quantitative analysis, considering only those options that were technically feasible at the time, offered potentially large release reductions, addressed different environmental media, and posed no process or worker safety problems. Projects designed to comply with several current or anticipated regulations were also included.

The workshop also addressed screening criteria to help assign priorities to the options and to list potential barriers and incentives for implementation and permitting concerns. The diverse viewpoints brought to these discussions helped guide subsequent project activities and reinforced the work group's desire to consider broader issues, such as multimedia release management consequences and the impact on future liability. The workshop was able to consider these strategic issues more comprehensively than either government or industry alone.

Ranking the Choices and Counting the Savings

In evaluating reduction options, the work group recognized that some of the options might transfer pollution from one medium to another. For example, upgrades of the drainage system and process water treatment plant would improve process water treatment and reduce air emissions, but would also produce more solid waste such as biosolids and spent activated carbon. Keeping soils out of sewers would reduce the amount of sludge in the API separator and thus allow for more on-site management of other solid wastes, reducing off-site disposal. Installing an electrostatic precipitator would reduce fluid catalytic cracking unit particulate emissions, but transfer the additional collected particulates to land disposal.

The work group agreed to consider the full pollution prevention hierarchy—source reduction, recycling, treatment, and safe disposal—as the basis for developing and evaluating all the release reduction options. The technology options identified and analyzed fit into the pollution prevention hierarchy. Less than half the options identified, however, qualified as true source reduction, and those that did managed only 20 percent of the pollutants (see Table 21.1). Had the options been limited to only source reduction, the scope of potential opportunities for reducing releases and improving environmental quality would have been unnecessarily restricted.

The project used a multidimensional ranking process—the analytical hierarchy process—in which weights were developed for all criteria used to rank options.[2] The process allowed the work group to assess the significance of and interactions between criteria. Criteria included risk reduction, cost factors, and technical characteristics.

TABLE 21.1 Amoco/EPA Pollution Prevention Project: Summary of Pollution Prevention Options

Option	Pollution prevention mode	Present value of capital (M$)	Net release reduction (t/yr)	Cost-effectiveness ($/t)
1. Reroute desalter water	Recycle	1,000	47.5	6,921
2. Improve desalter system	Undefined	Undefined	Undefined	Undefined
3a. Replace FCU cyclones	Recycle/disposal	8,300	222.3	13,628
3b. Install ESP at FCU	Disposal	9,100	401.0	8,935
4. Eliminate coker B/D pond	Source reduction	2,000	117.9	5,359
5a. Install secondary seals on gasoline tanks	Source reduction	212	430.6	209
5b. Install secondary seals on gasoline and distillate tanks	Source reduction	262	437.4	256
5c. Install secondary seals on all floating roof tanks	Source reduction	364	490.8	316
5d. Option 5c and floaters on fixed tanks	Source reduction	1,532	536.8	1,187
5e. Option 5d and secondary seals on fixed tanks	Source reduction	1,682	537.2	1,300
6. Decrease soils in drainage systems	Source reduction	213	480.8	422
7A. Blowdown system upgrade	Treatment	4,521	4,623.0	353
7B. Drainage system upgrade	Treatment	18,800	102.1	58,212
7C. Water treatment plant upgrade	Treatment	22,500	52.6	140,697
8. Modify sampling systems	Source reduction	67	57.2	473
9. Reduce barge loading emissions	Recycle	4,700	696.7	2,308
10. Sour water system improvements	Recycle/treatment	605	16.3	12,187
11a. Annual LDAR program @ 10,000 ppm	Source reduction	5	289.8	317
11b. Quarterly LDAR program @ 10,000 ppm	Source reduction	5	463.1	298
11c. Quarterly LDAR program @ 500 ppm	Source reduction	5	640.0	304

Note: All cash flows are discounted at 10 percent, 15-year project life. Capital spending for all projects is assumed to begin in 1991. Cost-effectiveness is based on annualized costs which include capital, operating and maintenance costs, depreciation, indirect costs, taxes, and insurance. ESP = electrostatic precipitator. FCU = fluid catalytic cracking unit. LDAR = leak detection and repair.

Although source reduction was usually more cost-effective than other choices, very few of the options considered would pay for themselves. If all the hydrocarbons recovered by source reduction could be collected and sold as gasoline, the gasoline cost would range from $0.23 to $4.23 per liter ($0.88 to $16 per gallon), with an average cost of $0.66 per liter ($2.50 per gallon). At the time of the project analysis, the average refinery price for gasoline was about $0.20 per liter ($0.75 per gallon). There was no economic incentive to recover these polluting materials.

Although Amoco and the EPA brought different perspectives to the discussions, each organization reached the same conclusions about which options would be most effective and which would be least. The driving forces in this priority setting were cost and relative risk reduction as measured by benzene exposure.

The project work group ranked control of marine loading losses as the most effective—although not the lowest-cost—option to reduce benzene exposure. A second tier of options included installing secondary seals on tanks, reducing soil intrusion into underground sewers, instituting a leak detection and repair program, and upgrading blowdown stacks. Amoco, the EPA, and the Commonwealth of Virginia jointly selected these five options. All five were also viewed as reasonably cost-effective pollution prevention projects. These five projects would prevent or capture almost 6800 t/yr of pollutant releases at a cost of about $550 per metric ton.

In contrast, compliance with current and anticipated regulations requires controls for eight refinery sources, resulting in release reductions of 6600 t/yr at an average cost of $2650 per metric ton. If the refinery were allowed to meet this same release reduction target by using its choices of sources and control techniques as identified by the project, it could save approximately $40 million per year.

Making Sense of the Experience

Public-private partnerships can be effectively used to leverage EPA resources for providing improved data needed to develop regulations. This project illustrates a possible approach to collecting data, assessing technologies, and characterizing a facility within an industry that took less time and EPA resources, but relied more on private support than is the current practice.

The Yorktown project produced a great deal of knowledge and understanding for both industry and government agencies. Although the refinery had to work within existing regulations, many pollution reduction measures did materialize out of the project. Improvements were made to the sewer system, stormwater system, and storage tanks; and process changes allowed the coker blowdown pond, a source of pollution, to be removed from service. Modifications to operating units are reducing benzene emissions, and a project to reduce barge loading emissions is awaiting issuance of final technical regulations.

Just as important, however, was the precedent of cooperation that the project established. With positive relationships founded on open-mindedness, cooperation, and mutual respect, industries and agencies can work together to develop a better understanding of environmental problems and more innovative, effective solutions.

Information generated through a facilitywide multimedia assessment is a necessary first step not only to developing a strategy to reduce these releases but also to exploring such implementation options as integrated permits. Industrial waste cost accounting systems that reflect total waste costs and EPA accounting systems that measure direct benefits of EPA activities must follow.

Data from this study show that the refinery can meet a release reduction goal more cost-effectively than the cost of achieving reductions prescribed by current regulatory or legislative requirements. In light of this finding, the EPA might evaluate options for setting a goal or target for reducing multimedia releases from a facility and then allow the facility to develop a strategy to meet the goal. This alternative strategy would allow the facility to meet the goal at a lower cost, include interim milestones, and be enforceable. This strategy would also make appropriate information publicly available to ensure that the reduction targets were met.

Data from this study show that an emissions inventory could be improved by measuring releases and developing new emission factors. The emissions inventory at the beginning of the project did not account for all potential releases to the environment. Some releases were excluded from reporting by the EPA (e.g., barge loading operations), other releases were not included because they were thought to be insignificant (e.g., blowdown stacks), some emissions were overestimated (e.g., the API separator), and other emissions were underestimated (e.g., coker quench pond). Jointly established sampling and analysis protocols could help improve data quality, so that reported values more accurately portray facility emissions.

The quality and utility of toxic release inventory data [as required by the Superfund Amendments and Reauthorization Act (SARA)] could be improved by

- Providing more inclusive estimates of facilitywide releases to all media

- Reporting groups of chemicals [for example, volatile organic compounds (VOCs)] rather than individual species, especially if these chemicals have similar structural, physical, and toxicological properties

- Reporting selected chemicals of concern for demonstrated human health or ecological impact separately

- Improving emission factors for estimating releases based on information developed in this project

The refinery is a major source of the area's VOC emissions. However, information on the potential adverse health effects of very low-level VOC exposure is rather limited.[3] Research is needed to better characterize health and ecological effects of VOCs for use in conducting risk assessments. This study could also build on efforts currently underway at the American Petroleum Institute and the Chemical Industry Institute of Toxicology, among others.

The EPA should also undertake research to develop indicators that measure impacts on the ecosystem of multimedia releases from industrial facilities. This project looked at several biomarkers that show promise as indicators of the health of aquatic environments. When these biomarkers were used at this facility and location, no significant impacts were found.

Lessons for Strategists

The Yorktown experience demonstrates the opportunities and pitfalls that can occur when government and industry work together. The opportunities are significant. The pitfalls are worth overcoming. All organizations—the EPA, the Commonwealth of Virginia, and Amoco—sought to develop and test innovative environmental management approaches that, unlike most traditional command-and-control approaches, consider risk reduction, address multimedia concerns, maximize environmental benefits, encourage efficient use of resources, and promote facility-specific implementation choices.

Although it will take time and patience to overcome decades of distrust, such joint government/industry efforts can result in more cost-effective environmental protection by providing the opportunity to share different viewpoints and skills. The enthusiasm of the people representing the many public and private organizations involved confirms the central premise of the project: Developing effective solutions to complex environmental management problems will take the best efforts of many partners in U.S. society.

References

1. Y. Cohen, D. T. Allen, D. N. Blewitt, and H. Klee, "A Multimedia Emissions Assessment of the Amoco Yorktown refinery," presented at the 84th Annual Air and Waste Management Association meeting, Vancouver, Canada, June 17, 1991.
2. T. L. Saaty, *Decision Making for Leaders*, RWS Publications, Pittsburgh, 1990.
3. J. E. Graham, ed., *Harnessing Science for Environmental Regulation*, Praeger, New York, 1991.

22

Dow's Energy/WRAP Contest: A 12-Year Energy and Waste Reduction Success Story*

Kenneth E. Nelson

President
KENTEC Incorporated
Baton Rouge, Louisiana

Summary

Keeping employees interested in saving energy and reducing waste is a constant challenge. Ideally, all employees should routinely find ways of improving operations as part of their normal jobs.

The Louisiana division of the Dow Chemical Company began an energy conservation program in 1981. It took the form of an annual contest. In 1983, the contest scope was expanded to include yield improvement, and in 1987, Dow's Waste Reduction Always Pays (WRAP) program was added. The contest has been enormously successful in achieving a continuous improvement attitude. In the "History and Results" section of this chapter, we will look at the contest history and summarize some of the results achieved. The "Why the Contest Works" section will

---

*Copyright Dow Chemical Company.

examine how the program operates, emphasizing what works and what does not.

History and Results

In the late 1970s and early 1980s, many companies jumped on the energy bandwagon and started energy conservation programs. But as energy prices dropped, most of these programs were phased out. The program at Dow's Louisiana division is an exception. Instead of disappearing, it continued to grow and gain strength. The tenth, eleventh, and twelfth years of the contest had the greatest participation in its history. In this section, we look at some of the results that have been achieved.

The Louisiana division

The Louisiana division of the Dow Chemical Company is located in Plaquemine, Louisiana, 10 miles from Baton Rouge, the state capital. It has about 2000 employees and over 20 plants making products such as ethylene, propylene, chlorine, caustic, polyethylenes, glycol ethers, chlorinated solvents, and many others. Internal steam and power needs are supplied via cogeneration, primarily by using gas turbine combined-cycle units.

1981—The first contest

In 1981, fuel gas prices were rising (and projected to rise even further), and like most companies, Dow's Louisiana division set up an energy conservation program. It was in the form of an annual Energy Contest, and a newly formed Energy Evaluation Committee took on the task of administering it. The contest was aimed at capital projects, and the requirements were simple. Projects had to meet three criteria:

- Projects had to be capitalized. Expensed projects and routine maintenance projects did not qualify.

- Projects had to cost less than $200,000. This simplified and expedited authorization.

- Projects had to have a *return on investment* (ROI) greater than 100 percent. Although this figure might seem high now, most companies had the same requirement in the early 1980s. It was unclear how long fuel prices would remain high, and the philosophy was to get our money back in a year or less.

The contest was aimed at engineers in the production plants, and it began without a lot of fanfare. It was announced, appropriate forms

TABLE 22.1   1982 Contest

Winners	Capital	ROI
27	$1.7 million	173%

were distributed, and a deadline was set. This occurred in late 1981. Since projects were to be done in 1982, the contest was called the 1982 Energy Contest.

Early results

A total of 39 projects were submitted. The Energy Evaluation Committee took on the chore of reviewing each entry. When reviews were completed, 27 projects survived. Half of the claimed heat savings did not exist! Still, the final results remained impressive. Results are shown in Table 22.1.

Once these 27 projects were uncovered, many people felt there could not be others with such high returns, but the contest was allowed to continue for another year. The 1983 Energy Contest was run in a similar manner and had even more spectacular results (Table 22.2).

Then the division general manager made some changes. First, he removed the $200,000 limit. "If there's a million-dollar project out there with a 100 percent ROI, we ought to be looking for it," he commented. Second, he expanded the scope of the contest to include yield savings. Today we would call this *waste reduction,* but at the time that phrase had not come into vogue. The 100 percent ROI cutoff was maintained, and he asked that projects be audited after they are installed. Again, the results were impressive (Table 22.3).

In the years since 1984, the ROI hurdle has been lowered, but the contest format has remained essentially the same. These are some of the rule modifications:

TABLE 22.2   1983 Contest

Winners	Capital	ROI
32	$2.2 million	340%

TABLE 22.3   1984 Contest

Winners	Capital	ROI
38	$4.0 million	208%

- Both capitalized and expensed projects are included.

- Projects must save at least $10,000 per year. This minimum requirement helps focus attention on large savings and allows recognition of people who improve operations without spending money.

- Maintenance cost reduction projects are included. Routine maintenance projects, however, do not qualify. A project must eliminate (or substantially reduce) a maintenance problem.

- In 1986, Dow formalized its U.S. area waste reduction program and assigned it the acronym *WRAP*, for Waste Reduction Always Pays. The WRAP program became a part of the contest in 1987 (the 1988 contest), which was renamed the Energy/WRAP Contest. WRAP projects do not require an ROI payback or minimum savings, but they must reduce waste. Adding emission control devices does not qualify.

- In the 1993 contest, work process improvements were added. This category encourages people working in service functions to look for better ways to do their jobs.

Ultimately, virtually all cost-saving projects qualify as long as they result in real savings greater than $10,000 per year.

Continuing results

Participation in the Energy/WRAP Contest continued to grow, and the 1990–1993 period has been exceptional, with over 100 winning projects each year. In 1993, a new record was set for participation. The results are tabulated in Tables 22.4, 22.5, and 22.6. Although there is no longer a capital limit, projects costing more than $2 million are not included in the tabulations. There have been 14 such projects over the past 11 years. Including them would inappropriately skew the data.

In Table 22.4, you will note that the ROI cutoff has been gradually lowered (until 1992). This does not mean that a lower quality of proj-

TABLE 22.4   Summary of All Winning Projects (Less than $2 Million) Having ROIs above the ROI Cutoff

	1982	1983	1984	1985	1986	1987
Winning ROI projects	27	32	38	59	60	90
Average ROI (%)	173	340	208	124	106	97
ROI cutoff (%)	100	100	100	50	40	30

	1988	1989	1990	1991	1992	1993
Winning ROI projects	94	64	115	108	109	140
Average ROI (%)	182	470	122	309	305	298
ROI cutoff (%)	30	30	30	30	50	50

TABLE 22.5   Summary of WRAP Projects (Less than $2 Million)

	1988	1989	1990	1991	1992	1993
WRAP projects with ROIs above cutoff*	23	22	37	35	35	39
WRAP projects with ROIs below cutoff	1	3	16	18	19	28
Total winning WRAP projects	24	25	53	53	54	67
Average ROI of WRAP projects (%)	200	106	108	97	111	82

*These projects are also included in the Table 22.4 and Table 22.6 summaries.

TABLE 22.6   Dollar Summary of Winning Contest Projects (Less than $2 Million) Having ROIs above the ROI Cutoff

	1982	1983	1984	1985	1986	1987
Cost, $million	1.7	2.2	4.0	7.1	7.1	10.6
Savings, $1000/yr						
Fuel gas	2,970	7,650	6,903	7,533	7,136	5,530
Yield and capital	83	−63	1,506	2,498	798	3,747
Maintenance	10	45	−59	187	357	2,206
Miscellaneous	0	0	0	0	0	19
Total annual savings, $1000/yr	3,063	7,632	8,350	10,218	8,291	11,502

	1988	1989	1990	1991	1992	1993
Cost, $million	9.3	7.5	13.1	8.6	6.4	9.1
Savings, $1000/yr						
Fuel gas	4,171	3,050	5,113	2,109	5,167	4,586
Yield and capital	13,368	32,735	8,656	17,909	11,645	20,311
Maintenance	583	1,121	1,675	2,358	2,947	2,756
Miscellaneous	−98	154	2,130	5,270	518	788
Total annual savings, $1000/yr	18,024	37,060	17,575	27,647	20,277	28,440

ect has been allowed. The fuel gas cost also dropped, and an energy-saving project with a 30 percent ROI in 1987 would have roughly a 100 percent ROI if evaluated at 1983 fuel prices. Total contest participation (including projects greater than $2 million) is shown graphically in Fig. 22.1.

The 1989 contest results show lower-than-normal participation. There is a good reason for this. In 1988, most plants were running at capacity and sold out. Every extra pound meant extra profit. The contest was downplayed that year so that the plants could focus their efforts on production reliability. Although there was virtually no publicity, participation remained excellent. And because many of the projects included incremental capacity increases (high-profit pounds), the average ROI for the 1989 contest was the highest ever achieved!

Although yield savings and waste reduction projects were included in earlier contests, formally including the WRAP program in 1988

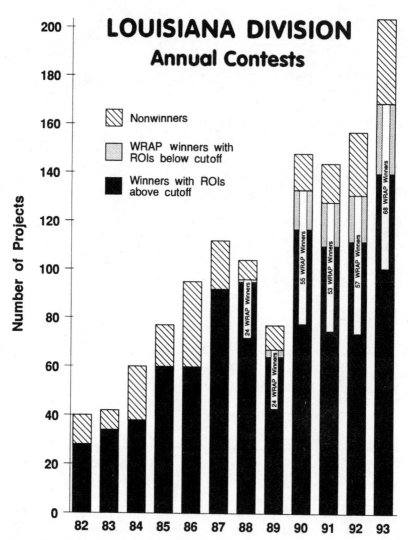

Figure 22.1   A 12-year history of Dow's Louisiana division contests.

gave the contest an added boost. Although WRAP projects do not require an ROI, all were above the 30 percent cutoff except for one project (it had a 25 percent ROI). Note the high average returns of WRAP projects shown in Table 22.5. Even though many WRAP projects have low or negative ROIs, the weighted-average ROI is usually over 100 percent.

Table 22.6 summarizes the cost and annual savings of contest projects. One of the curious results of requiring a $10,000 savings rather than a minimum cost has been the number of infinite-ROI projects (those requiring no capital or expense funds). The term *infinite ROI,* of course, is an oxymoron since we cannot have a return on investment if there is no investment.

But the phrase has caught on, and having such projects is a distinction. It has also allowed us to recognize the work of computer programmers, often the unsung heroes in a production plant. In the past 3 years, we have had 97 infinite-ROI projects which cumulatively save more than $16.6 million per year!

Audits

Audits occur after a project has been installed and running for a time. The purpose of audits is to compare each project's actual cost and performance with the contest values used to justify the project. An attempt is made to audit every project. Normally, only two or three members of the Energy Evaluation Committee attend an audit, which is very informal and discussion-oriented. Fluctuating fuel gas costs or product values are not important in audits. The important factors are actual project costs, the pounds or energy saved, and anything new that was learned. To date, 575 projects have been audited. Audit results are shown in Table 22.7.

On the average, projects do perform as expected. In fact, they perform a little better. Those high ROIs are real. The total audited savings for 575 projects are over $110 million per year!

Why the Contest Works

The author hopes that the reader is convinced that Dow's Louisiana division has an effective program—one that really gets results (and gets real results). In this section, we explain why the contest works. We take a closer look at some of the activities and values that have been a central part of the contest since its inception. We also include observations about the philosophy and logic behind the contest.

What works in Louisiana may not work everywhere, but many of these ideas should have wide application in other locations and in other industries.

TABLE 22.7   Audit Results

Total number of projects audited	575
Average submitted ROI (%)	202
Average actual ROI (%)	204

Sustained management support

Perhaps the most important reason for the ongoing emphasis of the Energy/WRAP Contest is the continuing support it receives from management. Of course, virtually all managers are in favor of saving energy, improving yields, reducing waste, etc. The key is to develop a program that is easy for them to support—one that does not conflict with other objectives or organizational responsibilities.

The way in which this was accomplished will become clearer in the following pages, but it was a concern from the start. When the program began in 1981, it did not require a new department, redeployment of people, or a multimillion-dollar budget—just a committee and a contest. Rather than set up separate functions, the approach became to use people in their existing positions and have project development and evaluation be a natural part of their jobs.

Organized Energy Evaluation Committee

Every program needs a focal point, and the Energy Evaluation Committee is the focal point for the Energy/WRAP Contest. This committee includes representatives from process engineering, economic evaluation, and various production areas.

The inclusion of a representative from economic evaluation is extremely important. The committee needs a person who knows the procedures for getting projects authorized and who works with managers in allocating the division's capital resources. Remember, the idea is to work within the system, taking advantage of the existing procedures and framework.

Started small

The program now includes both capitalized and expensed projects, there is no capital limit, and the scope of savings has been enlarged. But it did not begin that way. By starting small, committee members were able to work out the bugs and gain the confidence and trust of the plants. They did not need spectacular results the first few years, and it was not necessary to oversell the program.

Developed grass-roots support

Although the contest needs top management support to get it going and to sustain it, it is the grass-roots support at the plant level that gives it strength and vitality. Plant superintendents have found that the contest is an excellent vehicle for improving their processes. It is not a threat or a burden; it is a genuine asset.

The contest helps train young engineers, giving them a good sense of values. It provides a chance for them to demonstrate their technical, people, and communication skills. It creates peer pressure to come up with good projects and fosters a healthy, competitive spirit. Plant engineers work together, sharing ideas that help improve their plant. Although the contest was originally intended for engineers, an increasing number of nontechnical plant operators have generated ideas and submitted projects.

The contest does not conflict with plant priorities. In many plants, brainstorming sessions are used to uncover project ideas, and then the various ideas are prioritized and evaluated according to plant needs. Some of the strongest support comes from plant superintendents who were contest participants when they were production engineers. They understand the value of the program and make sure that their production engineers participate.

People with problems are looking for solutions

This may seem deceptively simple, but in many plants and organizations, people are not expected to look for solutions to their problems. Instead, they are expected to cope with them. In these organizations, those assigned the responsibility for solving problems are not close enough to fully understand or appreciate the magnitude, importance, or impact of those problems. In Louisiana, the Energy/WRAP Contest provides a simple mechanism and motivation for problem solving.

Kept paperwork simple

The contest entry forms have not changed much over the years. They ask for a minimum of information:

- Project description
- Summary of utility, yield, and cost savings (unit costs are given)
- ROI calculation (the formula is part of the form)
- Before-and-after sketches
- Waste reduction summary (required by WRAP program)

Only one copy is submitted (additional copies are made as needed), and it does not have to be typed. Minor corrections and updates are made on the original. One copy of the backup data and calculations is also required. It is filed with the original to facilitate follow-up when the project is audited. The contest is probably one of the few programs that actually receives compliments for its meager paperwork.

Reviewed projects

Project reviews have become an important part of the Energy/WRAP Contest. They work on the premise that nobody wants to install a project that does not work. Their purpose, then, is to ensure, as best practical, that projects will perform as expected. Committee members also recognize the importance of not surprising or embarrassing people, especially in front of their peers or supervisors.

Even before all projects are turned in, a tentative review schedule is set up. Letters are sent to all participants, documenting the time of their reviews and explaining what should be discussed. Typically, four or five members of the Energy Evaluation Committee are scheduled to attend a review. Each receives copies of the projects ahead of time. If they find errors or omissions or have any questions about the validity of a project, they are asked to contact the person(s) submitting the project before the review.

Throughout the review process, the focus is on evaluating projects, not people. When there are errors, the committee's job is to correct them, not to attack the people who made them. Having experienced people on the Energy Evaluation Committee is important. They understand the importance of working with people, helping them to explain and develop their ideas.

The tone during the reviews must be supportive. People submitting projects should leave feeling good about the review. If their project is "shot down," they should be thankful they were stopped from wasting time installing a project that was uneconomical or would not work. It is this sense of working together to a common end that must prevail.

In the early days of the contest, a few plant superintendents did not like the idea of "outsiders" coming into their plant and evaluating their projects. By maintaining the integrity of the reviews, requiring people to justify their projects, and by treating everyone equally, committee members gradually gained the confidence and support of these superintendents, who came to appreciate that the committee members were truly there to help, not to be an obstacle. As one superintendent put it, "You give us a hard time, but then again, you give everybody a hard time."

Held contest once a year

The contest is designed to fit with the division budgeting process, with October as the deadline for entries. Reviews are held in November and December, and results are finalized in January. This does not mean that projects cannot be worked on earlier. Reviews can be held at any time (at the request of the plant), and many of the entries—especially very high-ROI projects—are completed during the year and turned in for recognition.

Having a deadline has turned out to be an important feature of the contest. Some people just need deadlines to motivate them to action. Typically, more than 90 percent of all entries are submitted within a week of the deadline.

Contest did not control capital

Although the Energy/WRAP Contest has become an important tool for uncovering and prioritizing projects, neither the contest nor the Energy Evaluation Committee controls any capital. That function rightly remains in the hands of the division managers and is administered through the Economic Evaluation/Capital Planning department.

Control of capital is a sensitive issue. By not attempting to control part of the capital budget, those administering the contest have avoided some difficult and potentially conflicting situations. Instead of competing for funds, the contest has become an integral part of the budgeting process, helping to define capital needs and project timing. Many plants view the contest as a good way to get capital for projects because funds are normally made available to contest winners.

Winners received recognition—not cash!

After seeing the spectacular results of the contest, many people ask about the size of cash awards given to winners. Well, there are none— at least none that are directly related to the contest. Instead, a formal awards ceremony is held in which the division general manager presents all winners with engraved plaques.

We have avoided any type of cash awards for several reasons:

1. It is impossible to be fair. Who thought up the idea? Who worked on it? Who implemented it? How is the size of the award determined? Is a $10,000 project with a 1000 percent ROI better than a $1 million project with a 50 percent ROI? And what if the project does not work? Waiting until a project is installed and operating may take too long. The motivational benefits are lost, and personnel changes may occur that obscure who should receive cash awards.

2. Cash awards for contest projects can be demotivating. People with outstanding projects will feel they should have received more, and those who were unable to submit projects because of their particular work situation (e.g., assigned to a team building a new plant) may feel cheated.

3. In a plant situation, rewarding individuals with cash will quickly inhibit communications and stifle teamwork. People will be reluctant to share information and ideas, fearing that someone else will get the credit (and the cash).

4. Within any company, there are established procedures for monetarily rewarding people for good performance. Normally, each individual's supervisor has the primary input on the size of that reward, whether it is in the form of an annual salary increase or bonus. If a separate system is established, such as cash awards for contest winners, it literally competes with the boss for an employee's loyalty and tends to result in conflicting project priorities. The effect may be subtle, but the net outcome will be a loss of support from supervisors.

5. Perhaps the most important reason for avoiding cash awards for contest winners is that it implies that finding significant improvements is not part of the regular job—it is not what they are being paid for. They came up with good ideas, so they are being paid something extra. This is exactly the opposite of the message that should be sent. Employees should feel that continuous improvement is a normal and important part of their jobs. It is not something separate.

Although the contest does not give cash awards, this does not mean that people who submit good projects are not appropriately rewarded. But it is done by their own supervisors through normal channels. An individual's supervisor is in the best position to evaluate her or his contribution to improved plant operations (through the Energy/WRAP Contest or other vehicles) and to put it into the context of total job performance.

The author has spoken with many people from other companies about their methods for rewarding employees. Most companies that established cash award systems wish they had not. One company actually increased participation when it eliminated cash awards.

Worked through existing line organization

One of the advantages of having a committee administer the contest is that it does not create new levels of hierarchy or bureaucracy. It is necessary to work with existing departments and to follow existing procedures. This approach has worked well. There are no cross-purposes. Everyone is trying to accomplish the same thing, and in some cases the contest has helped people function more efficiently in their jobs.

The annual Energy/WRAP Contest has become an integral part of division procedures, adding to that grass-roots support discussed earlier. It is a part of the culture. It has not become a threat to anyone's job security.

Credit went to plants

Part of the Energy Evaluation Committee's strategy is to see that plants are recognized for participating in the contest and that the credit for projects goes to the people who thought them up and implemented them. Neither the contest nor the committee takes credit for savings. In fact, an important function of the committee is to find ways of appropriately recognizing good performance.

Educated and trained people

The educational and training benefits of the Energy/WRAP Contest are subtle, yet substantial. In addition to learning the technical design aspects of a project, many engineers are able to follow their project from the idea stage through preliminary evaluation, formal evaluation, authorization, engineering, construction, and start-up.

While developing and implementing projects, they learn how to get through the system. Along the way, they gain valuable experience in dealing with a variety of departments, disciplines, vendors, other engineers, supervisors, and operators. This develops technical ability, people skills, and self-confidence. There is no substitute for the positive feelings one gets after successfully bringing a project from the idea stage to full implementation.

Audits

Audit results were discussed in the "History and Results" section of this chapter and are an important and integral part of the contest. Audits give people a chance to brag about their successes. And if a project did not work out as expected, the Energy Evaluation Committee wants to know why. Remember, members of the committee were a part of the approval process, so if it did not work, they are partly to blame. But the issue here is not blame. The intent is to avoid making the same mistakes in the future. Documenting the reasons why projects fail to live up to their expectations helps accomplish that purpose (and keeps committee members humble).

Audits were begun back in 1983 at the request of the division's general manager (who was perhaps a bit skeptical about those high ROIs). Some plant resistance to audits was anticipated, but it never materialized. Instead, just the opposite occurred. Audits of project performance are strongly supported by plant superintendents. They encourage and endorse them.

Being able to document project performance has also given the Energy/WRAP Contest an enormous amount of credibility and has helped sustain its viability as an ongoing program.

No goals!

The Energy/WRAP Contest has never had numerical goals, such as number of participants, number of projects, dollars spent, or dollars saved. Instead, committee members recognized from the very beginning that the important thing to pursue was the mechanism—the process—by which projects are conceived, designed, and implemented. They needed to provide an organizational structure that functioned smoothly and made things happen. When an effective mechanism is in place, specific objectives and goals become unnecessary. Fine-tuning our mechanism is a never-ending job.

No gimmicks!

The contest has often been cited for the professional way in which it is run. Part of that professionalism includes a lack of gimmicks. The contest has never used slogans, posters, decals, bumper stickers, key chains, arm patches, caps, jackets, coffee cups, and so on to publicize saving energy and reducing waste.

There is nothing intrinsically wrong with these items, but they do little, if anything, to generate projects that save energy or reduce waste. Using a jacket for recognizing winners (instead of a plaque) is another matter, and it may work well as a motivating tool. But posters on the wall that say "Save Energy" or "Reduce Waste" do nothing to help accomplish their well-meaning intents.

Sometimes the gimmicks become the focal point of the program. A contest to design the best poster, e.g., can consume weeks of effort that might be better spent looking for projects.

No monthly newsletter

Monthly newsletters are helpful for recognizing people, generating ideas, and spreading information. But they can be extremely time-consuming, and missing an issue (or publishing a poor-quality issue) sends out the signal that the program is being deemphasized. Another problem with monthly newsletters is that after a few months, good ideas are essentially lost. People rarely take the time to read through back issues when they are faced with a new problem.

Gave people ideas (not projects)

There are very few technology breakthroughs. Nearly all new projects are applications of old principles. The Energy Evaluation Committee uses three highly effective means of communicating good project ideas:

1. Each year, a complete listing of all Energy/WRAP Contest projects is published. Both winners and nonwinners are included (note that they are not called *losers*). Each item in the list contains the name of the plant, the people submitting the project, a brief description of the

project, and the cost, savings, ROI, and waste reduced. The summary is organized by ROI. Project descriptions are worded to give an understanding of the principles involved without getting into detail.

The Energy/WRAP Project summary gets read by almost everyone involved with the contest. They want to know what others are doing and often find ideas applicable to their own plants—to submit in the next contest.

2. Several editions of a *Waste Elimination Idea Book* have been published, the most recent in 1993. It contains improvement ideas organized by subject (pumps, heat exchangers, distillation, compressors, etc.). Project ideas were taken from past contests and from work done at other Dow locations.

3. A Continuous Improvement Workshop is held every year or two. This is an intensive 2-day course aimed at giving people practical ideas for improving plant processes. It includes analytical, creative, and pragmatic approaches to finding cost-saving projects as well as specific information about various unit operations.

One session, entitled "How Winners Think," is based on interviews with the top 12 project generators in the division. Their methods for coming up with many good projects year after year are explained to workshop participants, who are challenged to apply the same insights and techniques in their own jobs.

The workshop concludes with a 2-hour session where attendees split into groups of four or five and analyze a hypothetical flow sheet, looking for problems and suggesting ways to improve the process. There are no "right" answers, and a wide variety of ideas are exchanged.

Participants then return to their plants and have the opportunity to use the knowledge gained. They are required to generate at least one new project. The group reconvenes 2 weeks later to discuss what each participant experienced when trying to apply the principles learned in the workshop. Response to the workshop has been excellent.

Conclusions

The program works! Hundreds of excellent projects have been implemented, and on the average they perform even better than expected. The program is supported by the division general manager, major managers, plant superintendents, plant engineers, and operators. It has become part of the division culture. Finding improvements is accepted as an important part of the job.

The contest format as practiced in Dow's Louisiana division provides an atmosphere where generating projects is fun—it is challenging. There is a minimum of red tape. Teamwork and cooperation within and between plants flourish, and there is a continually building momentum toward bigger and better projects with higher ROIs. In fact, the number of high-ROI projects submitted each year continues to be a source of amazement.

Addendum 1

It is important to recognize that not all waste reduction projects are cost-effective and that it takes time to develop waste reduction technology. Such technology is highly desirable, however, because it permanently reduces waste.

In the past, regulatory emphasis and mandates have been directed at waste treatment rather than waste reduction, because treatment technology is known and conversion of undesirable products is virtually complete.

Treatment devices, however, are usually very expensive, they are not cost-effective, and they incur ongoing operating expenses. Waste reduction options should be thoroughly exhausted before treatment technology is turned to. Recognize, however, that treatment is sometimes the best option.

It is hoped that requirements for further treatment devices will be limited to those areas where there is a clear and urgent need. This will enable our stretched and limited resources to be more effectively directed at reducing waste.

Addendum 2

Each year, we are faced with new demands, new regulations, new legislation, new challenges, and, of course, new problems. Amid all this activity, we never want to lose sight of one extremely important concept:

You reduce waste by installing projects that reduce waste.

You can state corporate positions, write detailed mission statements, outline ambitious goals, and prepare detailed plans, but

You reduce waste by installing projects that reduce waste.

You can give dynamic speeches, publicize inspiring slogans, develop lofty mottos, design award-winning posters, and distribute multicolored brochures, but

You reduce waste by installing projects that reduce waste.

You can establish high-level committees, appoint technical task forces, organize environmental teams, and restructure entire departments, but

You reduce waste by installing projects that reduce waste.

You can meet with environmentalists, negotiate with regulators, confer with politicians, talk to community leaders, and address the general public, but

You reduce waste by installing projects that reduce waste.

You can write letters to the newspaper, be interviewed on radio, appear on TV, speak at technical meetings, lecture at symposia, and even testify before Congress, but

You reduce waste by installing projects that reduce waste.

By now, the message should be clear. A successful waste reduction program must be more than talk. It must get results. It must raise employee awareness and concern about environmental issues to the point where specific ideas for projects to reduce waste are conceived, designed, and implemented.

For Dow's Louisiana division, the annual Energy/WRAP Contest has proved to be a highly effective and successful mechanism for accomplishing these objectives. Examples of projects and process modifications uncovered as a part of this program are given in Chap. 10 of this book. Readers are encouraged to explore similar applications in their own processes.

23

Southern California Edison's CTAP: Saving Energy and Reducing Pollution for Industrial Customers

Roger D. Sung, Ph.D.
Southern California Edison Company
Rosemead, California

Jimmy D. Kumana and Alan P. Rossiter, Ph.D.
Linnhoff March, Inc.
Houston, Texas

Introduction

From 1987 to 1989, Southern California Edison (SCE) conducted a pilot program to offer technical assistance to major industrial customers in energy efficiency improvement through process integration. Within this trial *customer technical assistance program* (CTAP), process integration studies were carried out for seven plants, representing a total load of 44 MW, and covering a broad range of industries in SCE's service territory as follows:

Industry/product	Number	SIC code
Paperboard mills	2	26
Brewery	1	20
Gypsum wallboard	1	32
Oil refinery	1	29
Glass containers	2	32

Although the original motivation was to improve energy efficiency, it became apparent, as the work proceeded, that reduced air emissions were an important side benefit. If all the recommended projects were implemented, then CO/CO_2 and SO_x emissions would be reduced by 22 percent, and NO_x emissions by 38 percent. The link between energy efficiency and emissions reduction is clearly shown in the results of these studies. In many cases, both benefits are attainable at a very attractive payback on the capital invested.

A follow-up survey in 1990 showed that there had been a high degree of implementation (or expected implementation), representing over 60 percent of the potential revenue improvement.

The CTAP illustrates how systematic process improvement methods can be used to help manufacturers save energy and reduce emissions. It is also a good example of cooperation between a utility company and its industrial customers to address matters of mutual concern. This chapter summarizes the results and implications of the work and then highlights the management approach taken and methodology used.

Background and Objectives

The prevailing trend in power demand in southern California has been an increasing residential and commercial load, but a declining industrial base load. To a large extent, the loss in load has resulted from cogeneration projects implemented in the wake of the PURPA regulations. However, some of the loss has been due to facilities simply being closed down because of the high cost of doing business in the area. As a result, Southern California Edison found itself facing a number of adverse business conditions involving

- Surplus generating capacity

- Continued loss of base-load industrial customers

- Continued deterioration of load profile caused by growth of the commercial/residential sector

In 1987, the research division of SCE began exploring innovative new concepts to reverse—or at least slow down—these trends. The chemical and related industries were specifically targeted, because they tend to be large base-load users of heat and power, operate continuously year-round, and have the capability to cogenerate if they choose to do so.

The conventional approach of offering rate relief to counter cogeneration projects was considered undesirable, for obvious reasons. Instead, it was reasoned, helping the customers achieve equivalent improvements in energy efficiency without resorting to cogeneration might accomplish the desired objectives for both parties.

The basic concept is quite simple: If the thermal efficiency of an industrial plant is improved through enhanced heat recovery, this reduces the size of the available heat sink, which in turn reduces the thermal match for cost-effective cogeneration. Furthermore, if the reduction in steam and/or fuel use were sufficiently large (say, 20 to 30 percent) and at a reasonably good payback (say, 2 to 3 years), in all likelihood the customer would implement the heat recovery projects in preference to cogeneration. There are several good reasons for this:

- Heat recovery using heat exchangers is a well-known technology, and people are comfortable with its application.

- There is little or no technical or economic risk.

- Projects can be implemented gradually, over time. Cash flow and capital requirements are more easily managed than they would be for large cogeneration projects.

The key to the success of this concept is that the heat recovery projects, as a group, must be able to compete effectively against cogeneration for management attention. Ideally they should be able to offer comparable savings with less risk and equal or shorter paybacks.

The research division required a common-denominator methodology that could meet these criteria across a broad range of industries and process technologies. It found the essential component of such a methodology in pinch analysis (see Part 2 of this book). In addition, expertise in specific items of equipment, operations, and relevant process technologies was brought to bear on individual studies as appropriate.

A pilot customer technical assistance program (CTAP) was initiated by SCE to develop and test the application of this concept in real business environments. Ultimately, a total of seven process improvement studies were undertaken between 1987 and 1989, to demonstrate and further refine the methodology. These seven customers were assisted:

Industry	Customers	Electric load, MW
Paper milling	2	14.1
Brewing	1	10.0
Gypsum wallboard	1	1.5
Glass containers	2	10.0
Oil refining	1	8.5
	7	44.1

Although the initial scope of work was restricted to improving thermal efficiency in the process, it was quickly recognized that reducing thermal load has a direct impact on the CO/CO_2, SO_x, and NO_x emissions that result from combustion. Projects would therefore have environmental benefits in addition to reducing energy costs. Moreover, other productivity objectives such as capacity debottlenecking, aqueous pollution abatement, waste minimization, and improved raw materials utilization are also important to the customer. Accordingly, the project objectives were selectively broadened to include some of these as well.

Methodology

The assistance provided for each plant typically consisted of developing a plant efficiency improvement plan involving a mix of energy reduction, capacity debottlenecking, and pollution abatement, by using pinch analysis.

The scope of services provided under the CTAP intentionally covered a fairly wide range—high, medium, and low—in order to establish the correlation between level of engineering effort and impact on customer response.

High-level services included making several site visits, developing a detailed understanding of the customer's process and specific needs, and providing a full technical and economic feasibility analysis of the project recommendations, including flow sheet drawings.

Medium-level services consisted of one or two site visits and making specific project recommendations, but without a detailed technical and economic feasibility analysis to support the recommendations.

Low-level services (scoping studies) consisted of setting theoretical energy targets and establishing the potential scope for energy savings, but without detailing how these savings could be achieved in terms of equipment needed and design modifications required.

The technical assistance was provided in the form of process improvement consulting services contracted for by SCE with outside consultants. For each customer, a consulting team was assembled that had the proper mix of skills to address the study objectives. Each consulting team had three to five members:

1. Representative from SCE research division

2. Representative from SCE customer service department (usually the major account executive)

3. Consultant with expertise in pinch analysis

4. Consultant with expertise in the customer's process technology

5. Consultant with expertise in equipment operations that dominate the customer's process

In some cases, two or more of the skills required (the last three items above) were possessed by the same individual, thus minimizing the number of people involved. It was considered critical that the consulting team be truly seen as experts by the customer. The pinch analysis consultant played the lead role, since pinch analysis was the common-denominator expertise applicable to all studies. The specific steps involved are detailed in Table 23.1. The members of the consulting team listed above who were responsible for each of the steps are also identified in Table 23.1.

Results

The project results are summarized in Table 23.2, together with their implementation status at the time of the customer survey carried out in late 1990. Only limited data are presented for the oil refinery study, because the work was restricted to scoping (i.e., identification of how much energy could be saved, without defining specific projects for implementation).

The projects identified include a large number of conventional heat integration schemes (i.e., recovery of waste heat for beneficial use). In addition, several improvements in plant steam systems were identified, together with some changes in plant operating conditions and process technologies that give rise to energy savings. Opportunities for using advanced electrical technologies, such as variable-speed drives, were also identified, along with load-shifting and peak load shedding options that are beneficial both to the plant operator and to the utility company. Monitoring and targeting (M&T), a methodology for tracking and improving plant performance by using graphical and statistical techniques, was identified as a potentially attractive management tool for the two glass plants. One potentially viable opportunity for gas turbine cogeneration was also identified (wallboard plant).

Reduction of air emissions

Table 23.3 provides estimates of the potential impact of the identified energy savings on air emissions. Any reduction in energy consumption will reduce combustion-related emissions and thus will help to mitigate air quality deterioration. This is particularly important in the Los Angeles basin, with its severe air quality problems.

There is a direct linear correlation between CO/CO_2 and SO_x emissions and the quantity of any given fuel being burned. The impact on NO_x emissions is harder to quantify, since it arises both from the fuel itself (fuel NO_x) and from the oxidation of nitrogen during combustion (thermal NO_x). Fuel NO_x emissions are directly proportional to the amount of fuel burned and its nitrogen content. Thermal NO_x emis-

TABLE 23.1   CTAP Methodology

Task 1 Screen SCE's customer database to identify candidate plants. Selection criteria included size of thermal and electrical bills, continuous or batch operation, and the consultant's prior experience in that industry regarding opportunity for cogeneration.	By 1*
Task 2 Contact candidate customers regarding their interest in participating, e.g., willingness to provide data and commit worker resources for interaction with the consulting team.	By 1, 2
Task 3 Preliminary visit to customer site to explain the CTAP, expected benefits, and detailed data requirements and schedule, and to develop a clear understanding of the customer's priorities.	By 1, 2, 3
Task 4 Preparation of a detailed work plan and budget, including recruitment of additional consulting expertise as needed.	By 3
Task 5 A 2- to 5-day site visit to collect process and equipment data and to develop rapport with customer's personnel.	By 3, 4, 5
Task 6 Preparation of process flow sheet drawing and performance of pinch analysis to establish energy targets and to identify potential process improvements, alternative heat recovery schemes, and the optimum plant utility system.	By 3, 4
Task 7 Review of equipment operations as well as reserve capacity analysis.	By 3, 4, or 5
Task 8 Site visit to review proposed recommendations and obtain customer feedback regarding feasibility and constraints.	By 1, 2, 3
Task 9 Revision of project recommendations as appropriate.	By 3, 4, 5
Task 10 Development of savings and capital cost estimates.	By 3, 4, 5
Task 11 Preparation of draft report and submission to SCE.	By 3
Task 12 Review of draft report by SCE.	By 1, 2
Task 13 Site visit for presentation of results, in person, to customer.	By 1, 2, 3
Task 14 Preparation and submission of final report.	By 3
Task 15 Customer follow-up 3 months and 6 months after study.	By 2

*See designations of consulting team members 1 through 5 in text.

sions must be estimated from empirical correlations for a given combination of boiler and burner type, fuel, and operating rate. Thermal NO_x emissions rise with increasing flame temperature in the combustion chamber. In general, it is possible to reduce the flame temperature at reduced operating rates by increasing the amount of excess air; however, the boiler efficiency suffers.

Other pollution abatement

At paper mill 1, wastewater discharge to the sewer was significantly reduced by one of the recommended projects.

At the brewery, the preparation of a heat and material balance revealed some avoidable raw material losses that had been overlooked

TABLE 23.2   Revenue Impact and Implementation Status

Customer	Potential revenue improvement ($1000/year)	Estimated capital ($ millions)	Simple payback (years)	Percent-age of revenue being implemented/ evaluated	Engineering detail of study
Paper mill 1	1,950	1.3	0.7	82	High
Brewery	6,420	3.0	0.5	99	High
Wallboard plant	3,430*	6.5	1.9	0	Medium
Oil refinery	2,150	Not eval'd.	n/a	0	Low
Glass plant 1	750	0.5	0.7	95	Medium
Glass plant 2	470	Not eval'd.	n/a	40	Medium
Paper mill 2	930	0.5	0.5	97	High
Total	16,100	11.8	0.7	60	

*Excluding potential 17 MW gas turbine.

TABLE 23.3   Energy Savings and Air Emission Abatement by Customer

Customer	Energy savings		Pollution abatement			Capacity increase (%)
	Electric (kW)	Fuel (MW)	CO_2/SO_2 (%)	NO_x (%)	BOD (t/yr)	
Paper mill 1	940	3.1	19	35	1100	0
Brewery	940	8.8	36	59	2700	40
Wallboard plant	−500*	2.2	8	15	0	30
Oil refinery	Not eval'd.	14.7	25	44	0	0
Glass plant 1	540	3.8	22	38	0	0
Glass plant 2	190	3.1	20	36	0	0
Paper mill 2	180	0.8	13	25	0	20
Total	2290	36.5	22†	39	3800	‡

*Excluding potential 17 MW gas turbine.
†CO_2 emissions would be reduced by approximately 168 t/day if all projects were implemented. The SO_2 emissions reduction cannot be estimated because the sulfur contents of fuels were not recorded.
‡Capacity increase is an alternative to energy savings.

by plant personnel. A few simple piping changes eliminated the loss, giving yield improvements valued at several hundred thousand dollars per year. At the same time, the wastes rejected by the process were correspondingly reduced. Since the "lost" material had previously been discharged into the sewer system, the changes also led to a reduction in the wastewater biological oxygen demand (BOD).

Additional benefits

There were several additional benefits derived from this work, quite apart from energy savings and emissions reductions, of which the most significant may be summarized as follows:

- Capacity debottlenecking (without requiring additional fuel firing and thus without creating additional combustion-related emissions)
- Assistance in defining investment priorities for industrial clients
- Improved customer relations and customer/load retention for SCE

Implementation Status

Data on project implementation status of the recommendations given to each facility were obtained by conducting telephone interviews with the appropriate person from plant staff. By mid-1990, projects representing approximately 60 percent of the potential revenue increase had been already implemented, were in the process of being implemented, or were under active consideration for implementation (see Table 23.2).

The record shows that customer response seems to correlate strongly with the degree of engineering assistance provided. Where detailed project recommendations were provided, incorporating a substantial engineering effort to establish technical and economic feasibility, the degree of implementation was very high. Where specific project recommendations were made but without engineering backup, the degree of implementation depended on how much the customer's staff and management had "bought into" the program. Where only energy targets were given, the degree of implementation was zero.

Conclusions

The environmental benefits of the pilot CTAP work could be significant for those customers who are facing compliance orders to meet increasingly stringent air emissions regulations. In general, the reduction in CO_2, SO_2, and NO_x emissions from the plant utility system is directly proportional to the fuel savings. Fuel savings are obtained at the same time, yielding a reduction in operating costs. This makes energy conservation using pinch analysis an economically attractive option for reducing pollution. In some cases, it may provide an alternative to installation of expensive end-of-pipe pollution control equipment (see also Chap. 18).

Initiatives based on this methodology offer a good way to promote both energy savings and pollution abatement. This approach is particularly applicable in areas like the Los Angeles basin, where air pollution problems are particularly severe. Programs of this type offer many potential benefits for the region as a whole, including not only pollution abatement and energy conservation, but also increased competitiveness of local industry, with consequent improvements in the local economy.

If all the recommended projects had been implemented, the total improvement in revenues would have been approximately $16,000,000 per year. Furthermore, CO_2 and SO_x emissions would have been reduced by 22 percent, and NO_x emissions by 38 percent. With an actual implementation rate of 60 percent, a large portion of these benefits will ultimately be obtained.

The pilot CTAP has provided some clear and tangible benefits to both customers and SCE. Experience with the seven studies has demonstrated that the basic concept is sound and that the methodology is workable. The degree of implementation appears to be strongly correlated to the engineering effort expended on the study and the customer's confidence in the study results.

The most interesting outcome was proof that pollution abatement need not be economically burdensome. On the contrary, with good process integration it can, in fact, be profitable.

Programs similar to SCE's CTAP have now been implemented by several other utility companies, and also by the Electric Power Research Institute (EPRI), Palo Alto, California.

Index

ABOUT THE EDITOR

Alan P. Rossiter is a principal project engineer with
Linnhoff March, Inc. in Houston, Texas, where he is responsible for conducting process integration studies for industrial clients, and for developing new methodologies and software for pollution prevention. The author of numerous technical articles, conference presentations, and published
reports, Dr. Rossiter has extensive experience in environmental abatement in a wide range of industries. He previously worked as a technical manager and consulting engineer with Imperial Chemical Industries.